Waste and Recycling

I0867165

As "business as usual" has become the mantra of today's world, it's unlikely to see a decrease in hazardous waste generated from greater economic growth. Written by renowned experts, the book suggests a solution, supported by theoretical arguments to this waste problem. The book discusses how main problems for waste management can be addressed through appropriate policies adopted by governments in OECD countries.

The book also raises thoughtful questions on how household waste management services should be privatized and who should pay for the disposal and recycling costs. It attempts to answer these questions. The book considers several factors hindering the first-best optimal outcome and highlights two crucial ones. It elaborates further with models and the solutions on how to overcome these obstacles.

The book covers not only traditional resource economics and waste management, but also the recent problem of electronic waste (E-waste) and illustrates, in detail, how the environments of developing countries are inevitably polluted even with the Basel Ban Amendment in place. The book proposes an alternative international trading regulation to address E-waste. This book will certainly appeal to industry decision-makers, policy-makers, and legislators.

Takayoshi Shinkuma has been with the Faculty of Economics of Kansai University since 2008. He taught at Gifu Shotoku Gakuen University and Tokyo University of Foreign Studies. He holds a Ph.D. in human and environmental studies from Kyoto University. Professor Shinkuma researches on environmental economics, specifically in waste management and recycling.

Shunsuke Managi is from Tohoku University and Institute for Global Environment Strategies. Previously, he taught at South Carolina State University, USA, Tokyo University of Agriculture and Technology, and Yokohama National University, Japan. He holds a Ph.D. in environmental and resource economics from University of Rhode Island, USA, and an M.Eng from Kyusyu University, Japan.

Routledge studies in ecological economics

1 **Sustainability Networks**
Cognitive tools for expert collaboration in social-ecological systems
Janne Hukkinen

2 **Drivers of Environmental Change in Uplands**
Aletta Bonn, Tim Allot, Klaus Hubaceck and Jon Stewart

3 **Resilience, Reciprocity and Ecological Economics**
Northwest coast sustainability
Ronald L. Trosper

4 **Environment and Employment**
A reconciliation
Philip Lawn

5 **Philosophical Basics of Ecology and Economy**
Malte Faber and Reiner Manstetten

6 **Carbon Responsibility and Embodied Emissions**
Theory and measurement
João F.D. Rodrigues, Alexandra P.S. Marques and Tiago M.D. Domingos

7 **Environmental Social Accounting Matrices**
Theory and applications
Pablo Martínez de Anguita and John E. Wagner

8 **Greening the Economy**
Integrating economics and ecology to make effective change
Bob Williams

9 **Sustainable Development**
Capabilities, needs, and well-being
Edited by Felix Rauschmayer, Ines Omann and Johannes Frühmann

10 **The Planet in 2050**
The Lund discourse of the future
Edited by Jill Jäger and Sarah Cornell

11 **Bioeconomics**
Edited by Mauro Bonaiuti

12 **Socioeconomic and Environmental Impacts on Agriculture in the New Europe**
Post-Communist transition and accession to the European Union
S. Serban Scrieciu

13 **Waste and Recycling**
Theory and empirics
Takayoshi Shinkuma and Shunsuke Managi

Waste and Recycling

Theory and empirics

Takayoshi Shinkuma and Shunsuke Managi

Routledge
Taylor & Francis Group
LONDON AND NEW YORK

First published 2011
by Routledge
2 Park Square, Milton Park, Abingdon, Oxon OX14 4RN

Simultaneously published in the USA and Canada
by Routledge
711 Third Avenue, New York, NY 10017

Routledge is an imprint of the Taylor & Francis Group, an informa business

First issued in paperback 2013

British Library Cataloguing in Publication Data
A catalogue record for this book is available from the British Library

Library of Congress Cataloging-in-Publication Data
Shinkuma, Takayoshi
Waste and recycling : theory and empirics / Takayoshi Shinkuma and Shunsuke Managi.
p. cm. – (Routledge studies in ecological economics ; 13)
Includes bibliographical references and index.
1. Refuse and refuse disposal–Case studies. 2. Recycling (Waste, etc.)–Case studies. 3. Waste products–Case studies. I. Managi, Shunsuke.
II. Title.
HD4482.S547 2011
363.7–dc22
2010041097

ISBN: 978-0-415-58985-7 (hbk)
ISBN: 978-0-415-70291-1 (pbk)
ISBN: 978-0-203-82801-4 (ebk)

Typeset in Times
by Wearset Ltd, Boldon, Tyne and Wear

To our parents

Contents

Figures

Tables

Acknowledgments

In writing this book, we have benefited from the wisdom and generosity of many. But we owe by far the greatest debt to Professor Yoshifusa Kitabatake of Kyoto University. He was the adviser of one of us (Shinkuma) and has always provided useful comments and encouragement to both of us.

We are also much obliged to Professor Eiji Hosoda of Keio University, who has been a constant source of inspiration. Professor Keisaku Higashida of Kwansei Gakuin University, Dr. Terazono of the National Institute for Environmental Studies (NIES), Assistant Professor Shinsuke Murakami of The University of Tokyo, Dr. Aya Yoshida of NIES, Dr. Rie Suzuki of NIES, Dr. Michikazu Kojima of the Institute of Developing Economies, Kenji Takeuchi of Kobe University, Dr. Yasuhiko Hotta of the Institute of Global Environmental Strategies (IGES), Dr. Satoshi Kojima of IGES, Associate Professor Masashi Yamamoto of Toyama University, Assistant Professor Daisuke Ichinose of Tohoku University of Community Service and Science, Professor Anthony Chiu of De La Salle University, and Dr. Amit Jain of IRG Systems South Asia Pvt. Ltd. provided enormous inspiration, encouragement, and full cooperation throughout the writing of this book. We are very grateful to them.

We thank Dr. N.T.M. Huong, a coauthor with Shinkuma in an earlier work, who not only provided insightful comments but also invested her energies in our scholarly development. We also wish to thank Yumi Skrzypczak, a professional English editor who has checked all the works of Shinkuma. It has been our great pleasure to work with the two of them. Appreciation is also due to participants in various conferences, workshops, and seminars for their comments and suggestions.

Funding has been received from the Environmental Economics and Policy Studies section of the Ministry of Environment, Japan (MOE), and a Grant-in-Aid for Scientific Research from the Japanese Ministry of Education, Culture, Sports, Science and Technology (MEXT, 21330060). We express our sincere thanks to these agencies. The results and conclusions of this research do not necessarily represent the views of these funding agencies.

We owe a great debt to Lam Yongling, our production editor at Routledge.

Finally, our gratitude goes out to our wives, Katsue and Keiko, and to our parents, for their encouragement and support.

This monograph contains some revised versions of previously published papers. Permission to reproduce materials published in the following papers was granted by Elsevier: Shinkuma, T. (2007), "Reconsideration of an advance disposal fee policy for end-of-life durable goods," *Journal of Environmental Economics and Management* 53: 110–121; Shinkuma, T. and N.T.M. Huong (2009), "The flow of E-waste material in the Asian region and a reconsideration of international trade policies on E-waste," *Environmental Impact Assessment Review* 29: 25–31; and Shinkuma, T. and S. Managi (2010), "On the effectiveness of a license scheme for E-waste recycling: The challenge of China and India," *Environmental Impact Assessment Review* 30: 262–267. Permission to reproduce materials published in the following papers was also granted by Springer: Shinkuma, T. (2003), "On the second-best policy of household's waste recycling," *Environmental and Resource Economics* 24: 77–95.

1 Introduction

1.1 What is waste?

When waste produced by people is not properly managed, serious environmental consequences can follow, such as human sickness and injury, pollution of the air and water resources, and destruction of amenity values. What exactly, then, is this "waste" that people have been talking about? The trading, transporting, treatment, and disposal of a thing are subject to the waste-related laws if it is identified as waste. Especially in the case of waste that is hazardous, the above activities are usually restricted to registered (or licensed) waste management firms. Because of these legal ramifications, there has been much controversy in OECD countries over the definition of waste.

There is no universally accepted definition of waste. In European Union (EU) member countries, waste is defined by the Council Directive 91/156/EEC as "any substance or object in the categories set out in Annex I which the holder discards or intends [to discard] or is required to discard." The notion of "discarding" embraces two types of operations: "disposal operations" listed in Annex IIA of the Directive and "recovery operations" listed in Annex IIB of the Directive. Recycling activities are, therefore, subject to the law. As the definition of waste given in the Basel Convention[1] also includes scrap to be recycled, the trading of waste for recycling must be controlled under the law.

Alternatively, waste can be defined as a thing whose price is negative. On the basis of this definition, a good is defined as a thing whose price is positive. This is the definition of waste that has been adopted in Japan in the Waste Disposal and Public Cleansing Law enacted in 1970. Under this definition of waste, a large number of recycling activities are not bound by the restrictions of the law, even if an improper method of recycling is adopted. For this reason the definition of waste adopted in Japan has been criticized as a serious defect in Japan's legislation.

Despite the criticisms leveled at this definition, it is the definition of waste that we have in mind throughout this book. The major reason is that the essential characteristics of the waste market stem from the negative price of waste, as we shall see below.

1.2 Economic development and waste generation

Does greater economic growth always hurt the environment? The Environmental Kuznets Curve (EKC) postulates an inverse U-shaped relationship between a specific measure of environmental pollution and per capita income levels. Income elasticity of environmental degradation turns from positive at lower levels of per capita income to negative at higher levels. Is the relationship of income and degradation linear or an inverted U-shape empirically? The answer, as so often happens, depends on pollutants (see Grossman and Krueger, 1995; Brock and Taylor, 2006; Managi, 2006; Managi *et al.*, 2009; and Tsurumi and Managi, 2010a, 2010b for a summary and discussion of the empirical literature).[2] In general, the EKC relationships are more likely to hold for short-term and local-impact pollutants than for pollutants with more global, indirect, and long-term impacts. Significant EKCs exist for soil degradation, for deforestation, for local air pollutants such as CO, NOx, sulfur dioxide, and suspended particulate matter, and for urban air concentrations. They reach a peak at lower income levels than the peak for global environmental indicators or some greenhouse gases and the depletion of fossil fuels.

The volume of per capita household waste has increased substantially due to the economic growth around the world. As regards municipal waste, empirical literature supporting the EKC is scarce (Kinnaman, 2009). Using macroeconomic data from 30 OECD member countries over 20 years, Johnstone and Labonne (2004) reject the hypothesis and show that the relationships between income and municipal waste generation are positive and linear. Figure 1.1 shows the cross-section data in relation to 21 countries in 2005.

The absence or weakness of evidence for EKC in terms of waste generation implies that even a mature developed country may continue to suffer from waste disposal problems and must keep making an effort to decrease the amount of waste generated. Even though the volume of waste has been reduced by compressing, incinerating, and recycling it, the depletion of final disposal sites continues to be a serious problem due to the resistance to new landfills or incinerators in several OECD countries. Figure 1.2 and Figure 1.3 show the capacity of municipal and industrial waste disposal sites in Japan, respectively.

1.3 The waste market

Competition in a waste market leads to a different result from competition in a conventional goods market. To begin with, let us consider a goods market. Because a particular good is valuable for both sides in a market—for supply (producers) and for demand (consumers)—competition in the market raises the quality of goods and pushes down the production cost. With regard to the goods, the market has a driving force toward high quality and low price. To see this, suppose that a producer of a good lowers the quality of that good. The purchaser of that good is concerned about the quality and will almost certainly become aware of the low quality in the course of its consumption. Thus, information

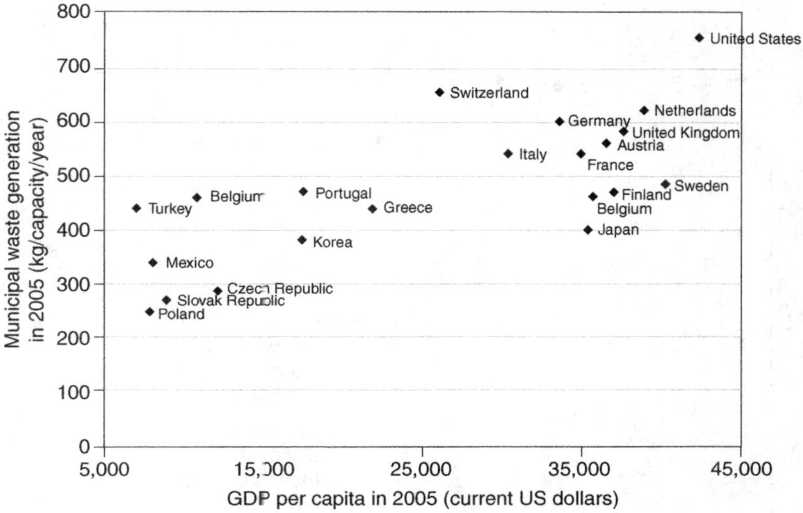

Figure 1.1 The relationship between GDP per capita and municipal waste generation (source: municipal waste generation information is obtained from IMF, *World Economic Outlook Database*, April 2010. GDP per capita information is obtained from *OECD Environmental Data Compendium* 2008).

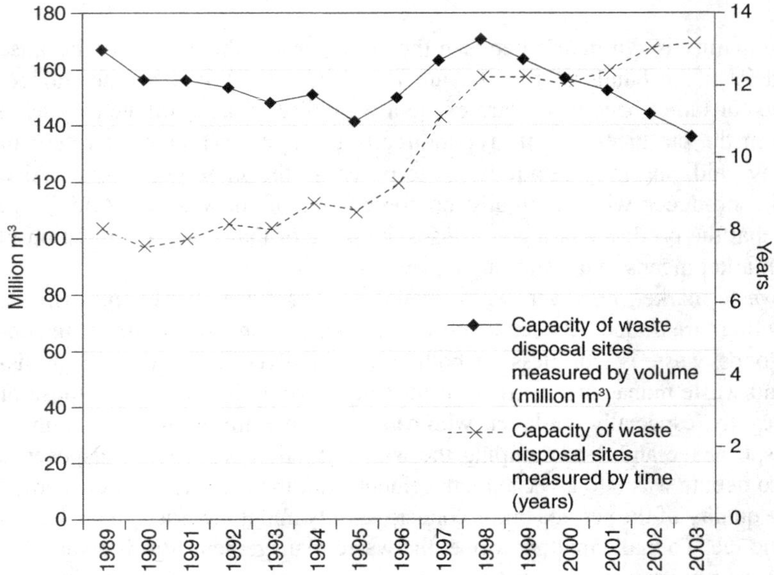

Figure 1.2 Capacity of waste disposal sites in Japan (municipal waste) (source: Ministry of Environment, Japan).

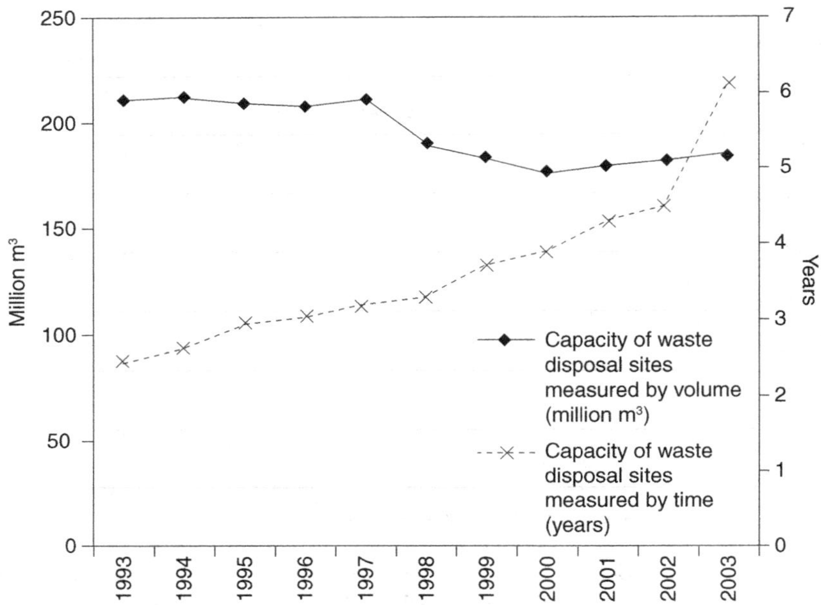

Figure 1.3 Capacity of waste disposal sites in Japan (industrial waste) (source: Ministry of Environment, Japan).

about the quality is symmetric between the producer and the consumer, because the good is in the hands of the consumer and its quality can be ascertained. When the consumer becomes aware of the low quality of the good, he may even claim from the producer directly (or indirectly through the retailer) a return of the money paid, and may decide never to purchase the same good again. As a result, the producer will eventually be forced out of the market. And if we suppose that the producer of a good raises the price of that good, competition in a goods market means that a similar story would unfold.

In a waste market, however, waste management services (including disposal and recycling) are traded between disposers and waste management firms. In contrast to goods, waste is worthless for both parties involved in the trade. Both disposers and waste management firms intrinsically have an incentive to dispose of waste illegally, especially the latter, who can make an extra profit by receiving a waste disposal fee and then dumping the waste. In addition, when a disposer is obliged to depute waste to a waste management firm, the disposer is unconcerned about the quality of the service but is concerned only about the price he or she has to pay: the fee. To make matters worse, the waste management firm is aware that the disposer is unconcerned. What is more, even if an eco-friendly disposer were to know how his waste was to be treated and where it was to be disposed of, it is quite difficult for him to know the quality of the waste treatment service, because the waste is already in somebody else's hands. As a result, there is asymmetric information between disposers and waste management firms about the quality of the service. This is why, as Hosoda (1999) points out, competition in the waste market will fail to raise the quality of the waste treatment service and can only push down the price. Consequently, waste gets treated with a minimum level of technology and the environment gets seriously damaged. And the most environmentally unsound treatment method is illegal disposal.

The possibility of illegal disposal is one of the most important aspects that must be taken into account when evaluating various types of policies throughout this book. In Chapter 2 we begin with industrial waste rather than household waste; it is in the industrial waste market that we find the unique characteristics of the waste management market described above. While household waste management has been provided by municipalities as a public service, industrial waste has been traded in a competitive market and treated by private waste management firms. In Chapter 2 we shall consider an optimal policy when both disposers and waste management firms have an incentive to dump waste. Even though the main theme of this book is how household waste management services should be privatized, an analysis of the industrial waste market will be rewardingly instructive.

1.4 EPR and privatization of household waste management services

Once we have studied the industrial waste market in Chapter 2, we shall consider household waste in Chapter 3. Household waste has traditionally been treated by municipalities. As long as municipalities are responsible for it, however, any signals of the cost associated with disposal and recycling cannot be transferred to either consumers or producers. The result is a market equilibrium in which there is greater production and consumption of goods than is consistent with the social optimal level. Therefore, to reduce the volume of household waste and conserve natural resources by promoting the recycling of household waste, responsibility for waste management needs to be transferred from municipalities to consumers or to producers.

A natural question is, then, who should pay the disposal and recycling costs: consumers or producers? In other words, what kind of policy is better able to address the problem, a disposal fee (DF) policy or a policy based on extended producer responsibility (EPR)? EPR is an environmental policy approach in which a producer's responsibility for a product is extended to the post-consumption stage of a product's life cycle. Under a simple setting we can state that EPR policies are superior to a DF policy. This is based on the following very simple logic: first, as shown in Chapters 3 and 4, in the first-best economy where there is no possibility of illegal disposal, a DF policy and EPR policies are equivalent; second, a DF policy does provide incentives for illegal disposal activities, whereas EPR policies do not; therefore, a DF policy is inferior to EPR policies.

1.5 The main objectives of this book

We have two objectives in compiling this book. The first is to reexamine EPR policies under more realistic conditions. For this purpose, we consider two cases. First of all, in Chapter 5 we consider the case where there are two sources of distortion: illegal disposal, and a transaction cost associated with a subsidy for recycling. It is shown in that chapter that the social optimum cannot be achieved when there is more than one source of distortion. Three policies are candidates for the second-best policy, two of them being EPR policies and one a DF policy. Second, we consider the case where the end-of-life goods are durable goods. This case is worth examining because an advance disposal fee (ADF) policy, which is one of the EPR policies, has been adopted especially for end-of-life durable goods in many OECD countries. So in Chapter 6 we examine the ADF policy and ascertain that it cannot achieve the social optimum and must be amended.

Our second objective in this book is to consider how the international trade for reuse and recycling should be organized or regulated. Recycling in a developed country is not closed. This is true especially in the case of end-of-life durable goods. Since the 1990s end-of-life electronic and electrical devices such as personal computers and home appliances, which are called E-waste, have been exported from developed to developing countries, and improper metal recovery processes from E-waste have seriously polluted the environment in various developing countries, but especially in China, India, and Pakistan. The background to, and the problems associated with, the trade of E-waste are overviewed in Chapter 7.

The material flow of E-waste depends on the policies that are adopted in the exporting and importing countries. Such policies also affect the level of damage caused to the environment and to social welfare through international trade. In Chapter 8 we investigate the effects that the introduction of an EPR policy in a developed country can have on international trade, the environment, and social welfare. We show that adopting an EPR policy in a developed country cannot be *globally* optimal. In addition, adoption of an EPR policy will necessarily decrease the social welfare in a developing country and increase the environmental damage in a developing country, seen from the perspective of the social optimum.

In Chapter 9, on the basis of data collected in our own field studies, we investigate how policies by importing countries affect the material flow of E-waste. Developing countries have adopted various types of policies with regard to the importation of E-waste. These policies are based on the Basel Convention, which acknowledges the right to ban the importation of hazardous waste. Enforcement of the Basel Convention regulations is unfortunately weak, with the result that the environment in importing countries, especially in China, India, and Pakistan, has been seriously polluted. The weakness of this enforcement has spurred member states to amend the Convention. That amendment, known as the Basel Ban Amendment, prohibits developed countries from exporting hazardous waste to developing countries. While the Basel Convention is a regulation whose adoption was led by importing countries, the Basel Ban Amendment is one whose adoption was led by exporting countries.

In Chapter 10 we investigate the effectiveness of a policy for discouraging improper recycling, i.e., a licensing system for E-waste recycling that has been adopted in China and India. We then evaluate the Basel Ban Amendment by comparing its advantages with its disadvantages. Finally, we propose an alternative trading regulation scheme. The recommended trading scheme is an extended license system that involves not only importing countries but also exporting countries.

The first objective of this book is connected with an important issue for developing countries that are intending to establish domestic recycling systems. The second objective is significant for both developed and developing countries that have established domestic recycling systems without taking the trade of waste into consideration. Reuse and recycling is not closed in each country. We hope that this problem is investigated further, especially by means of economic studies, and that a better solution can be found in the near future.

2 The industrial waste market

2.1 Introduction

As we have just seen in Section 1.3, in a market for waste management we are confronted with three characteristics: asymmetric information, heterogeneity of waste management firms (WMFs), and noncompliance. Waste is worthless both for disposers and WMFs. Disposers of waste are unconcerned about, and unaware of, how their waste is treated and how it damages the environment (i.e., asymmetric information). Competitiveness in the waste management market has not resulted in improving the quality of waste management but in lowering waste management fees. As a result, waste has been treated with a minimum level of technology and the environment has been seriously damaged. The lower level of technology makes it easier for firms to enter the waste management market. Indeed, various types of WMFs operate in the market (i.e., the heterogeneity of WMFs). Although the regulators have made efforts at improvement by setting a standard for how waste must be treated, the lack of concern among disposers has encouraged noncompliance by WMFs. In addition, disposers may themselves dump waste.

To address these problems, governments in OECD countries have introduced various kinds of policy devices, including penalties for illegal courses of action, a licensing system, and a *manifest* requirement system. In the next section we examine these policies and derive the optimal policy designs.

2.2 An optimal policy design: licensing, a fine on illegal disposal, and a *manifest* requirement[1]

Let us consider an optimal policy design with licensing, a fine on illegal disposal, and a *manifest* requirement. The policy examined in this section represents a currently prevailing (industrial) waste management policy that has been adopted in several OECD countries. In the RCRA (Resource Conservation and Recovery Act) Permit Program set up in the US, the owners or operators of hazardous waste treatment, storage, or disposal facilities are required to apply for RCRA permits. Those who generate waste must not pass on their hazardous waste to unauthorized treatment, storage, or disposal facilities. In the EU, operators of industrial and waste installations covered by the Integrated Pollution Prevention and Control

Directive are required to obtain an authorization (environmental permit) from relevant authorities in EU countries. There, too, those who generate waste are prohibited from contracting with unauthorized WMFs.

In our model we consider a competitive waste management market consisting of homogeneous waste disposers, two types of WMFs, and the government. It is assumed that disposers do not have any technology for waste treatment.

A technology level is represented by some level of environmental care[2] that is taken for one unit of waste. The extent to which the environment is damaged by the treatment of one unit of waste depends on the actual environmental care C that is taken in dealing with the unit of waste. The environmental damage resulting from the one unit of waste is represented by $E(C)$. It is assumed that $E'(C) < 0$ and $E''(C) > 0$. Let θ stand for the cost of one unit of environmental care.

In general, the unit cost of environmental care differs among WMFs. A technically inferior firm should pay more to acquire a waste treatment technology than a technically superior one. It also depends on financial accessibility, which differs from firm to firm.

The two types of WMFs differ from one another in regard to parameter θ. Without loss of generality, we assume that $\theta_1 < \theta_2$. The number of type i WMFs is denoted by n_i. The volume of waste treated by each WMF is fixed at 1. In addition, it is assumed that the volume of waste emitted by a disposer is also 1 and the number of disposers is A. We assume that the government knows the volume of waste of each disposer, the number of disposers, the capacity of each WMF,[3] and the number of WMFs for each type. It is the type of WMFs (which WMF belongs to which type) that the government cannot know.

In addition, we assume that[4] $n_1 > A$. It is socially optimal that only the cost-efficient, or type 1, WMFs operate in the market.

We can characterize the social optimality as follows. First of all, the social cost can be written as $A(\theta_1 C_1 + E(C_1))$. The government minimizes the social cost. The first-order condition (FOC) for the socially optimal environmental care C_1 can be represented by

$$\theta_1 + E'(C_1) = 0. \tag{2.1}$$

The government is supposed to attain the social optimum represented by (2.1) by means of a licensing system that is a standard-based regulation. The standard for environmental care is denoted by $\hat{C}$. Any WMFs that propose to operate can apply for the license.

The government is confronted with three problems: noncompliance of WMFs with the standard, asymmetric information, and the dumping of waste by disposers. The second problem means that the government needs to induce the inefficient, or type 2, WMFs to exit the market even if it cannot identify which WMFs are type 2.

Let us deal with the first problem, the noncompliance of WMFs with the standard. To focus on this problem, suppose that the government succeeds in getting type 2 WMFs to exit the market, or, to put it another way, that only type 1

WMFs operate. However, even type 1 WMFs may not comply with the standard by, for example, skipping some processes required by the standard or by using less costly but more harmful waste-treatment methods prohibited by the law. If x_1 stands for the noncompliance rate, the actual environmental care C_1 is calculated as $C_1 = \hat{C}(1 - x_1)$. Let us then introduce a fine on noncompliance. When a type 1 WMF does not comply with the required standard $\hat{C}$ and its noncompliance is detected, the government can impose a fine (denoted by z) on the firm. The noncompliance rate is denoted by x_1. The probability of detection is represented by r, which depends on noncompliance rate x_1. It is assumed that $r'(x_1) > 0$, $r''(x_1) > 0$, and $r(1) \leq 1$. A brief sketch of the licensing policy is shown in Figure 2.1.

The profit of a WMF (type 1) can be expressed as

$$\Pi_1 = P - \theta_1 \hat{C}(1 - x_1) - r(x_1)z,$$

where P denotes the fee for waste disposal or treatment. The firm decides on noncompliance rate x_1 so that Π_1 can be maximized. The FOC for the optimal noncompliance rate x_1 can be represented by

$$\theta_1 \hat{C} - r'(x_1^*)z = 0. \tag{2.2}$$

The government must decide on a standard $\hat{C}$ such that the *ex-post* environmental care can be equal to the socially optimal level. In this case, the government must choose a standard $\hat{C}$ that satisfies the following equation:

$$\hat{C}(1 - x_1^*) = C_1(\theta_1), \tag{2.3}$$

where $c(\theta_1)$ and x_1^* are defined implicitly by (2.1) and (2.2) respectively.

We can easily see that the social optimum can be attained if, for a given fine z, the government sets the standard $\hat{C}$ such that x_1^* and $\hat{C}$ satisfy both (2.2) and (2.3). Figure. 2.2 illustrates the socially optimal standard and fine. If the fine z is set at a sufficiently high level, we have two candidates or two curves representing (2.2)

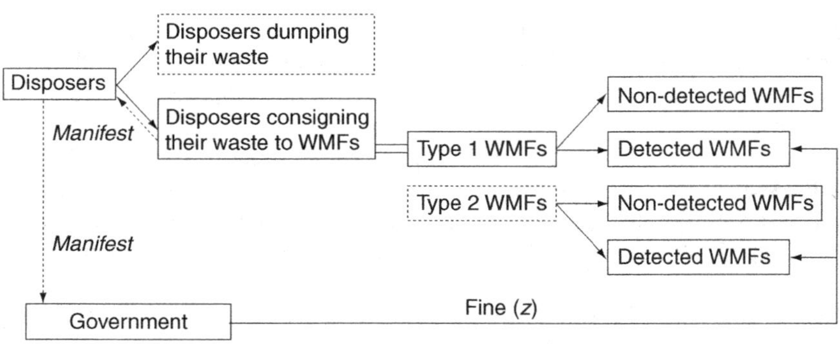

Figure 2.1 The licensing policy.

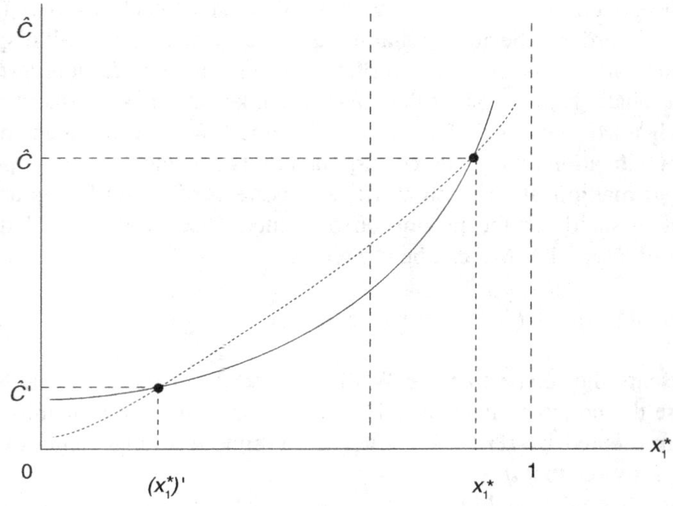

Figure 2.2 The optimal licensing policy. The solid curves and the dotted curves represent (2.3) and (2.2) respectively.

and (2.3) crossing each other. So we can be indifferent in regard to them because either can attain the social optimum. One of the two candidates is a combination of a high standard and a high noncompliance rate, and the other combines a low standard with a low noncompliance rate.

The second problem confronted by the government is that it must induce type 2 WMFs to exit the market, even if it cannot know for certain which WMFs are type 2. To guarantee that only type 1 WMFs operate in the market, the following equality and inequality must hold:

$$P - \{\theta_1 \hat{C}(1 - x_1^*) + r(x_1^*)z\} = 0 \tag{2.4}$$

$$P - \{\theta_2 \hat{C}(1 - x_2^*) + r(x_2^*)z\} < 0. \tag{2.5}$$

Inequality (2.5) implies that the profit of a type 2 firm is negative if it takes out a license. However, the competitive equilibrium ensures that this must hold, because it follows from $\theta_1 < \theta_2$ that $\{\theta_1\hat{C}(1 - x_1^*) + r(x_1^*)z\} < \{\theta_2\hat{C}(1 - x_2^*) + r(x_2^*)z\}$. For this reason, type 2 WMFs do not choose to apply for licenses. In addition, the assumption that $n_1 > A$ makes the equilibrium profit of each type 1 WMF zero. Obviously, the equilibrium waste disposal fee is $P = \{\theta_1\hat{C}(1 - x_1^*) + r(x_1^*)z\}$. The last problem to be addressed is that disposers themselves might dispose of their waste illegally. So we need an additional policy device; a *manifest* requirement. When the disposer consigns waste to a WMF, the government obliges the disposer to receive a verification letter (*manifest*) from the WMF and submit it to the government.

Note that the government knows the volume of waste of each disposer (i.e., one). Therefore, in order to be able to dump the waste by themselves, all disposers must ask a WMF to issue a false verification letter (a *pseudo manifest*) in exchange for a rebate. Note also that the government knows the waste-treatment capacity of each WMF (i.e., one). Therefore, each type 1 WMF can issue at most the amount of verification letters that corresponds to its capacity. From the point of view of profit maximization, each WMF will issue verification letters at the maximum. Let w stand for the proportion of issued false verification letters. Then the profit of a type 1 WMF can be written as

$$\Pi_1 = (1 - w_1)P + qw_1 - \theta_1\hat{C}(1 - w_1)(1 - x_1) - r(x_1 + w_1 - w_1x_1)z, \tag{2.6}$$

where q represents the rebate that the WMF receives from the disposer. Note that in this case the noncompliance rate is $x_1 + w_1 - w_1x_1$ and the probability of detection is represented by $r(x_1 + w_1 - w_1x_1)$. By simple arrangements of the FOCs we can easily see that $q = P$.

Now suppose that a disposal firm asks a type 1 WMF to issue a *pseudo manifest* and intends to dump the waste itself. In such a case the marginal benefit of a false verification for the disposer is P. However, the disposer needs to pay the rebate $q = P$ to the type 1 WMF for the latter to issue a false verification letter. Then the *marginal net benefit* of the disposer vanishes. This result can be stated as the following proposition.

Proposition 2.1. *If the government knows the amount of waste of each disposer and the waste-treatment capacity of each WMF, no fine needs to be imposed on the disposer in a system that requires a manifest.*

It has been considered that the *manifest* requirement can deter noncompliance by WMFs. However, according to the argument presented above, it discourages noncompliance behavior by disposers rather than noncompliance by WMFs, because the former must pay a rebate that is equivalent to the price for waste treatment by licensed WMFs.

Therefore, if the government chooses an appropriate level of fine and an appropriate standard and introduces a *manifest* requirement system, all three problems (noncompliance of WMFs with the standard, asymmetric information, and dumping of waste by disposers) can be addressed. As a result, the social optimum can be attained in a competitive market.

2.3 An alternative optimal policy: a deposit/refund system

It is difficult to set the fine for a particular non-compliant behavior at an extremely high level, because the penalties imposed for illegal dumping have to be weighed and balanced against those imposed for murder, theft, property damage, and so on—in other words, they need to match the gravity of the non-compliant behavior.

Alternatively, if we implement a deposit/refund system, we can substitute a tax and a subsidy for a fine. Under a deposit/refund system, economic agents who can take illegal actions must pay a fee in advance (the deposit), and then if they take legal actions they will be subsidized (i.e., the refund will be returned). As a result of a deposit/refund system, illegal actions will necessarily entail the loss of money. The central idea of a deposit/refund system is subsidizing a good course of action rather than directly imposing a tax or a fine on a bad course of action such as illegal disposal. One important advantage of a deposit/refund system is that it can impose a tax or a fine effectively (indirectly, in the form of a fee) on illegal and hidden actions that cannot be taxed or fined directly.

In this case, there are two sources of illegal courses of action: disposers and WMFs. So we need to revise the deposit/refund system, as shown in Figure 2.3. Under the revised deposit/refund system, a disposer must pay a deposit denoted by t per unit of waste. It can be considered a disposer tax. If the disposer consigns the waste to a WMF and it is verified by the *manifest* system, the disposer can receive a refund (denoted by s_1) that cannot be received otherwise. The cost to the disposer is represented by

$$P(1-y)+t-(1-y)s_1,$$

where y represents the amount of waste dumped by the disposer. It is easily shown that each disposer would choose to consign waste to a WMF rather than dump it by himself if $t = s_1 = P$. Note that the *ex-post* payment of a disposer is always P for any y.

The revised deposit/refund system also needs to prevent the noncompliance of WMFs. The noncompliance can be precluded simply by subsidizing compliance, or not being detected. The government gives a subsidy (denoted by s_2) to WMFs if they are not found to be non-compliant. Then the profit of each type 1 WMF is represented by

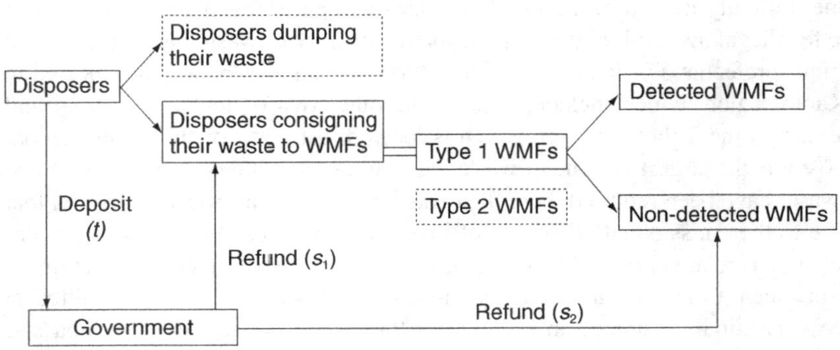

Figure 2.3 The deposit/refund system.

$$\Pi_1 = P - \theta_1\hat{C}(1 - x_1^*) + (1 - r(x_1^*))s_2.$$

The FOC for the optimal noncompliance rate x_1 can be represented by

$$\theta_1\hat{C} - r'(x_1^*)s_2 = 0.$$

Therefore, the subsidy for compliance is substitutable for the fine imposed on noncompliance. It follows that the optimal set of $(\hat{C}, s_2)$ is equivalent to that under the fine-based scheme discussed in Section 2.2: $(\hat{C}, z)$.

2.4 Illegal disposal of industrial waste

We choose to examine the case in Japan for empirical application because it is unique in that it summarizes policy questions also raised in other developed countries. Japan's basic law for waste management was enacted in 1970. The most important feature of the law is licensing. Waste management services are restricted to licensed WMFs, and operating without a license is prohibited by the law. If a licensed WMF is caught violating the regulations (for example, through illegal disposal), a penalty is imposed on the WMF and its license can be suspended for several years. A *manifest* system is also included in the policy for industrial waste management in Japan.

In addition, to reinforce the licensing system, liability has been extended to disposers by making them legally obliged to consign their waste to licensed WMFs. If a disposer violates this consignment obligation and the violation is detected, that disposer is penalized.

Since 1970 the law has been revised several times, and the penalties have been increased. At present, the maximum penalty that can possibly be imposed for illegal disposal is the equivalent of about $100,000 for individuals and $1,000,000 for corporations. Table 2.1 gives a brief history of changes to the law. Other penalties not included in the table are those for illegal burning of waste, reconsignment of waste to other WMFs, transporting waste for the purpose of illegal disposal, and issuing a *pseudo manifest*.

One difficulty in controlling illegal disposal is the transboundary movement of waste for illegal disposal. Figure 2.4 shows the quantity of waste illegally disposed of in four prefectures in Japan. Two prefectures (Chiba and Ibaragi) are located in the Kanto region, which includes Tokyo; the other two prefectures (Miyagi and Iwate) are in the Tohoku region, which is located just north of the Kanto region. Tokyo emits the largest volume of waste, but few cases of illegal disposal of industrial waste have been reported there. It seems logical to dump waste in an area that is close to the emission site (because of lower transport costs) and whose population density is relatively small (because of less probability of detection).[5] The figure confirms such a logical transboundary movement of waste. The waste emitted in Tokyo was initially dumped in prefectures that are closer to Tokyo: Chiba and Ibaragi. As a result of more stringent patrolling of those two prefectures, however, the waste was taken farther afield: to the two prefectures in the Tohoku region.

Table 2.1 Penalties in waste control in Japan

(1) Penalties for illegal disposal of industrial waste

Revision year	*Maximum possible fine*	*Maximum possible imprisonment period*
1970	$500	–
1976	$3,000	6 months
1991	$10,000 for hazardous waste $5,000 for other waste	1 year for hazardous waste 6 months for other waste
1997	$100,000 for individuals (for all waste) $1,000,000 for corporations	3 years (for all waste)
2000	$100,000 for individuals $1,000,000 for corporations	5 years

(2) Penalties for operating without a license

Revision year	*Maximum possible fine*	*Maximum possible imprisonment period*
1970	$1,000	1 year
1976	$5,000	1 year
1991	$30,000	3 years
1997	$100,000	3 years
2000	$100,000	5 years
2005	$100,000 for individuals $1,000,000 for corporations	5 years

(3) Penalties for violation of the consignment obligation

Revision year	*Maximum possible fine*	*Maximum possible imprisonment period*
1976	$3,000	6 months
1991	$10,000	1 year
1997	$30,000	1 year
2000	$100,000	5 years
2005	$100,000 for individuals $1,000,000 for corporations	5 years

Source: This table is a reprint of Table 1 in Shinkuma and Managi (2010b).

Note
1 US dollar = 100 yen (average exchange rate between June 2008 and May 2009).

Figure 2.5 shows recent trends in the total amount of industrial waste illegally disposed of, measured both by the total quantity dumped and the number of dumping incidents. Note that the quantity does not refer to the amount of industrial waste that was illegally *dumped* in the reported year but to the amount of illegally dumped waste that was *found* in that year. Therefore, we need to be aware of a time lag between illegal dumping and its discovery. For example, the sharp increase in quantity in 2003 is due to an incident in Gifu Prefecture in

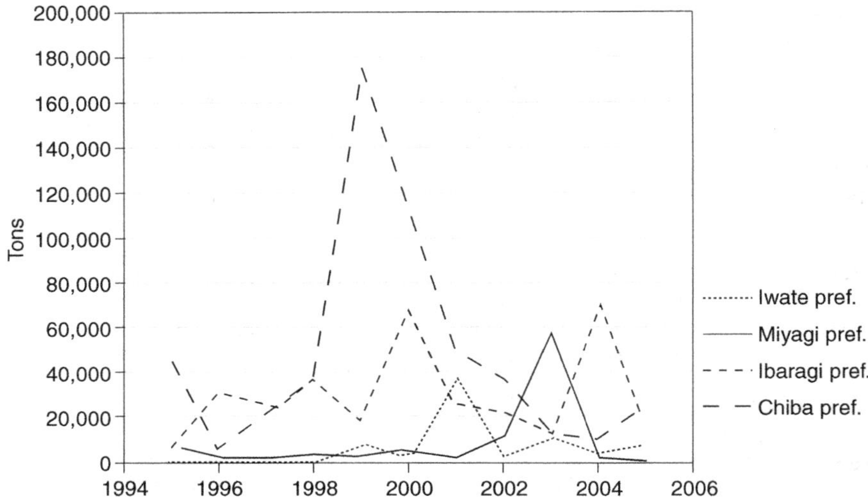

Figure 2.4 Transboundary movement of industrial waste for illegal disposal in Japan (source: Ministry of Environment, Japan).

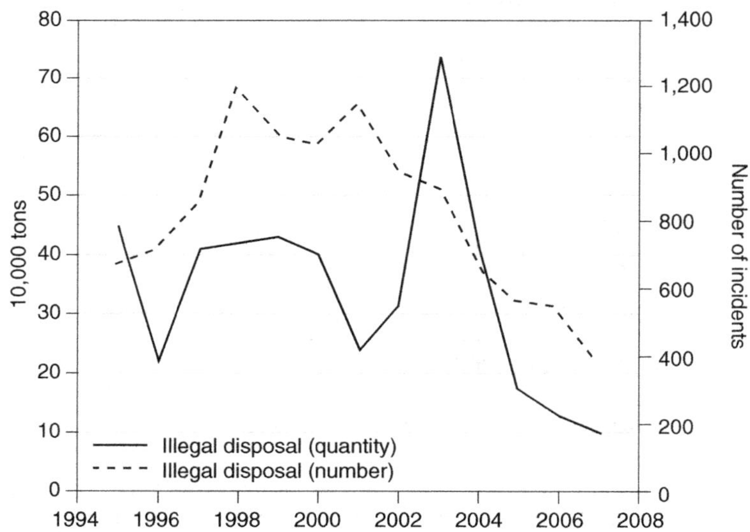

Figure 2.5 Illegal disposal of industrial waste in Japan (source: Ministry of Environment, Japan).

which a huge amount of industrial waste (567,000 tons) was found to have been illegally disposed of. This amount of waste was not dumped in 2003 but during the previous three or four years. Taking this time lag into account, we can state that illegal disposal of industrial waste has been decreasing due to increasing penalties for such disposal (see Table 2.1).

However, there are several points in the law that need to be amended. Figure 2.6 shows who dumped how much industrial waste in 2008. Disposers dumped about half the total amount. Even non-licensed WMFs have operated and disposed of waste illegally, even though the law prohibits operating without a license. The figure shows that there still exists a black market for illegal disposal.

According to the theory presented in Section 2.3, neither disposers nor unlicensed WMFs would have any incentive to dispose of waste illegally if the *manifest* system is functioning well. Note that the *manifest* system can work only if the government knows the total amount of waste and the waste-treatment capacity of each WMF (see Proposition 2.1). In practice, however, it is quite difficult for the government to know who is emitting waste and how much and the kind of waste that must be disposed of. To collect this kind of information, an integrated control might be needed over the material flow, from the production and sale of goods to their disposal.

Although Figure 2.5 shows a tendency for illegal disposal of industrial waste to go down in the years covered by the figure, in recent years improper waste treatment has been on the increase (see Figure 2.7). Ishiwata (2002) points out the possibility that *pseudo* recycling activities might weaken the control of the law and as a result promote improper treatment of waste. Improper waste treatment associated with *pseudo* recycling is possible because of the definition of waste. Note that in Japan waste is defined as a thing whose price is negative (see Section 1.1). If the nominal price of a waste is positive, it is considered to be a good or material to be

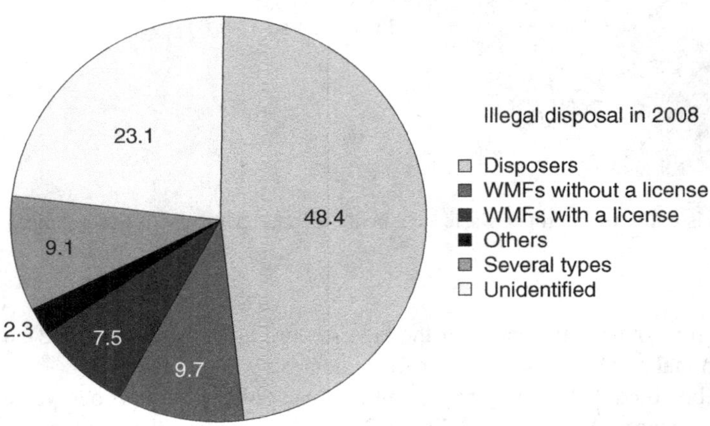

Figure 2.6 Who dumped industrial waste in Japan? (source: Ministry of Environment, Japan).

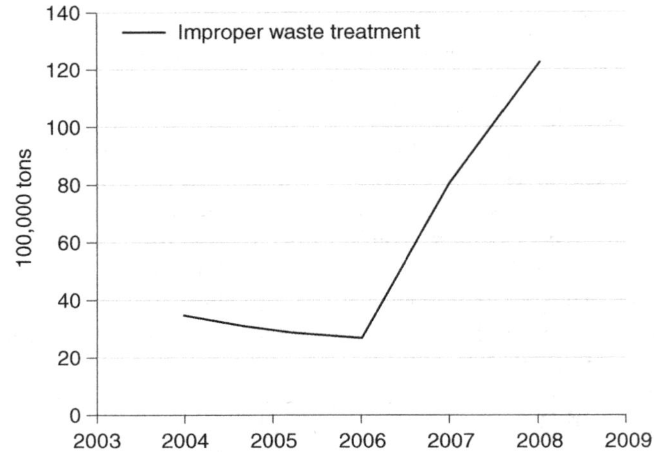

Figure 2.7 Improper waste treatment for industrial waste (source: Ministry of Environment, Japan).

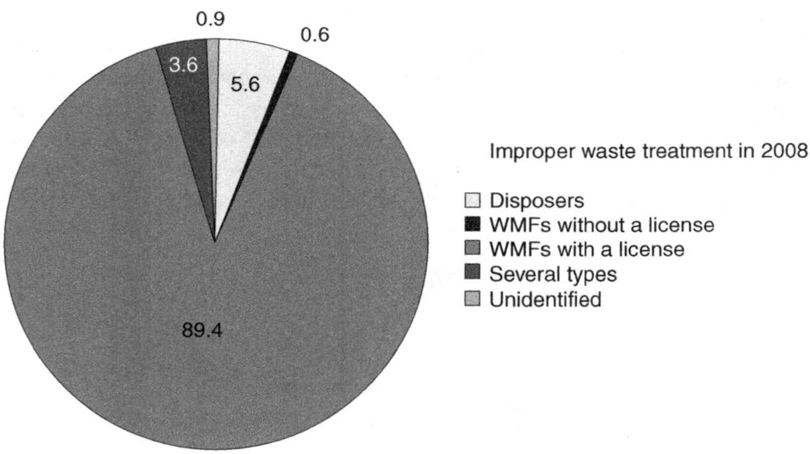

Figure 2.8 Who is guilty of improper waste treatment? (source: Ministry of Environment, Japan).

recycled and it falls outside the scope of the law. According to Ishiwata (2002), a simple trick can make the nominal price of waste positive.[6]

It must also be noted that almost 90 percent of those guilty of "improper waste treatment" are licensed WMFs (see Figure 2.8). This implies that "improper waste treatment" is concealed by legal activities such as recycling. The definition of waste used in Japan should be amended.

3 Creating the market for household waste and extended producer responsibility

3.1 Introduction

Municipalities have traditionally held the responsibility for undertaking treatment of household waste. Even though municipalities reduce the volume of waste finally disposed of through compression, incineration, and recycling methods, the volume of final disposal waste has been increasing because consumption has been increasing steadily. In several OECD countries the depletion of final disposal sites has become a serious problem as the result of opposition to new landfills or incinerators (i.e., the Not In My Back Yard (NIMBY) problem). In the next section we provide some empirical evidence that contradicts the Environmental Kuznets Curve (EKC) in the case of municipal waste.

As long as municipalities are responsible for household waste disposal, information about the costs involved in waste disposal and recycling does not get through to consumers and producers, and this results in greater production and consumption of goods than is consistent with the social optimal level. To reduce waste generation, the responsibility for waste management must be transferred from municipalities to consumers or producers.

The question arises: Who should pay the disposal and recycling costs—consumers, or producers? In other words, which is the better policy to address the problem, the disposal fee (DF) policy, or a policy based on extended producer responsibility (EPR)? EPR is an environmental policy approach in which a producer's responsibility for a product is extended to the post-consumption stage of a product's life cycle. There are several policies that are consistent with EPR (for example, a product take-back requirement, a deposit-refund system, an upstream combination tax/subsidy, or an advance disposal fee).

In recent years EPR-based policies have been introduced in many countries in preference to the DF policy. In Section 3.3 we consider the advantages of an EPR policy over the DF policy. Then we move on to the explanation for each EPR-based policy (Section 3.4). The final section provides several examples of EPR-based policies.

3.2 The EKC for municipal waste

In this section we provide simple econometric results that make use of 31 countries' national-level data on municipal waste generated from 1995 to 2002, obtained from Eurostat's data sources. This indicator tells us the amount of municipal waste generated. It consists of waste collected by or on behalf of municipal authorities and disposed of through the waste management system. The bulk of this waste stream is from households, though similar wastes from sources such as commerce, offices, and public institutions are included. For areas not covered by a municipal waste scheme estimates have been made of the amount of waste generated. The quantity of waste generated is expressed in kg per capita and per year. Economic data on GDP per capita in the unit of constant 2000 US$ is taken from world development indicators.

We present parametric results using ordinary least squares (OLS) and a fixed effects panel model. First, we estimate the OLS following the environmental quantity model, represented by:

$$\ln(waste_{it}) = \alpha_0 + \alpha_1 \ln(GDPpercapita_{it}) + \varepsilon_{it}$$

$$\ln(waste_{it}) = \alpha_0 + \alpha_1 \ln(GDPpercapita_{it}) + \alpha_2 [\ln(GDPpercapita_{it})]^2 + \varepsilon_{it}$$

$$\ln(waste_{it}) = \alpha_0 + \alpha_1 \ln(GDPpercapita_{it}) + \alpha_2 [\ln(GDPpercapita_{it})]^2 + \alpha_3 [\ln(GDPpercapita_{it})]^3 + \varepsilon_{it}.$$

Second, we estimate the fixed effect model, represented by:

$$\ln(waste_{it}) = \alpha_0 + \alpha_1 \ln(GDPpercapita_{it}) + \varepsilon_{it} + \gamma_i$$

$$\ln(waste_{it}) = \alpha_0 + \alpha_1 \ln(GDPpercapita_{it}) + \alpha_2 [\ln(GDPpercapita_{it})]^2 + \varepsilon_{it} + \gamma_i$$

$$\ln(waste_{it}) = \alpha_0 + \alpha_1 \ln(GDPpercapita_{it}) + \alpha_2 [\ln(GDPpercapita_{it})]^2 + \alpha_3 [\ln(GDPpercapita_{it})]^3 + \varepsilon_{it} + \gamma_i$$

$$\ln(waste_{it}) = \alpha_0 + \alpha_1 \ln(GDPpercapita_{it}) + \varepsilon_{it} + \gamma_i + \gamma_t$$

$$\ln(waste_{it}) = \alpha_0 + \alpha_1 \ln(GDPpercapita_{it}) + \alpha_2 [\ln(GDPpercapita_{it})]^2 + \varepsilon_{it} + \gamma_i + \gamma_t$$

$$\ln(waste_{it}) = \alpha_0 + \alpha_1 \ln(GDPpercapita_{it}) + \alpha_2 [\ln(GDPpercapita_{it})]^2 + \alpha_3 [\ln(GDPpercapita_{it})]^3 + \varepsilon_{it} + \gamma_i + \gamma_t,$$

where variables show data for country *i* and year *t*

The results of the waste generation equation are presented in Tables 3.1 and 3.2. The results in OLS show increasing trend over income on average. When time trend

Table 3.1 Environmental Kuznets Curve: time effect excluded

	OLS 1	*OLS 2*	*OLS 3*	*Fixed 1*	*Random 1*	*Fixed 2*	*Random 2*	*Fixed 3*	*Random 3*
ln(GDP per capita)	0.185*** (16.68)	–0.898*** (–3.93)	–19.257*** (–6.72)	0.241*** (8.18)	0.220*** (9.34)	–1.527*** (–5.34)	–1.341*** (–4.99)	–5.593* (–1.77)	–4.692 (–1.57)
(ln(GDP per capita))^2	–	0.059*** (4.74)	2.095*** (6.61)			0.100*** (6.22)	0.087*** (5.83)	0.552 (1.57)	0.461 (1.39)
(ln(GDP per capita))^3	–	–	–0.075*** (–6.42)					–0.017 (–1.29)	–0.014 (–1.13)
Constant	4.434*** (42.34)	9.341***	64.132*** (7.47)	3.904*** (14.07)	4.101*** (18.25)	11.586*** (9.17)	10.987*** (9.13)	23.640** (2.51)	20.922** (2.35)
Time effect	–	–		No	No	No	No	No	No
Hausman (chi^2)				1.40		7.27**		3.27	
Sample	433	433	433	433	433	433			

Note
* Significant at the 10% level.
** Significant at the 5% level.
*** Significant at the 1% level. Values in parentheses are t-values.

Table 3.2 Environmental Kuznets Curve: time effect included

	Fixed 4	*Random 4*	*Fixed 5*	*Random 5*	*Fixed 6*	*Random 6*
ln(GDP per capita)	0.122** (2.12)	0.163*** (5.02)	−1.655*** (−5.19)	−1.273*** (−4.65)	−5.167 (−1.62)	−5.705* (−1.85)
(ln(GDP per capita))^2			0.111*** (5.66)	0.082*** (5.28)	0.502 (1.42)	0.576* (1.69)
(ln(GDP per capita))^3					−0.014 (−1.10)	−0.018 (−1.45)
Constant	4.980*** (9.41)	4.597*** (15.20)	11.842*** (9.00)	10.822*** (8.91)	22.261** (2.34)	23.985** (2.61)
Time effect	Yes	Yes	Yes	Yes	Yes	Yes
Hausman (chi^2)	0.79		6.11		10.23	
Sample	433		433		433	

Note
* Significant at the 10% level.
** Significant at the 5% level.
*** Significant at the 1% level. Values in parentheses are t-values.

is not included in the specifications, panel model estimates show the fixed effect model for quadratic terms specification and the random effect model for cubic terms specification using the Hausman test. When time trend terms are included, the random effect model is selected both for the quadratic and cubic terms specifications. In all of the OLS, and the fixed effects and random effects models, there is a strong tendency for increasing trend over income to be higher. We cannot find any significant cases of an inverted U-shape at all. Among the three variables of GDP, the positive terms of linear and quadratic are most statistically significant. However, it might be that the EKC turning point exists at some specific income level. As some EKC studies indicate, in addition, the appropriateness of functional forms is problematic (e.g., Dasgupta *et al.*, 2002). We therefore intend to solve this problem using multivariable semi-parametric analysis. Semi-parametric regression analysis relaxes the assumption of linearity and typically substitutes the weaker assumption that the average value of the response is a smooth function of the predictors.

We use generalized additive models (Hastie and Tibshirani, 1990). We use a cubic spline smoothing iteratively to minimize the partial residuals, which are the residuals after removing the influence of the other variables in the model. In this iteration, the estimation loop stops when the model fit cannot be improved. In this model, a Bayesian approach is used to derive standard errors and confidence intervals. We include country dummy and year dummy to take into consideration individual and time fixed effects.

Figure 3.1 shows the predicted contribution to waste from GDP per capita. The central curve represents the estimated result. The upper and lower curves correspond to upper and lower 95 percent confidence intervals to account for heteroskedasticity and unrestricted correlation over time between observations

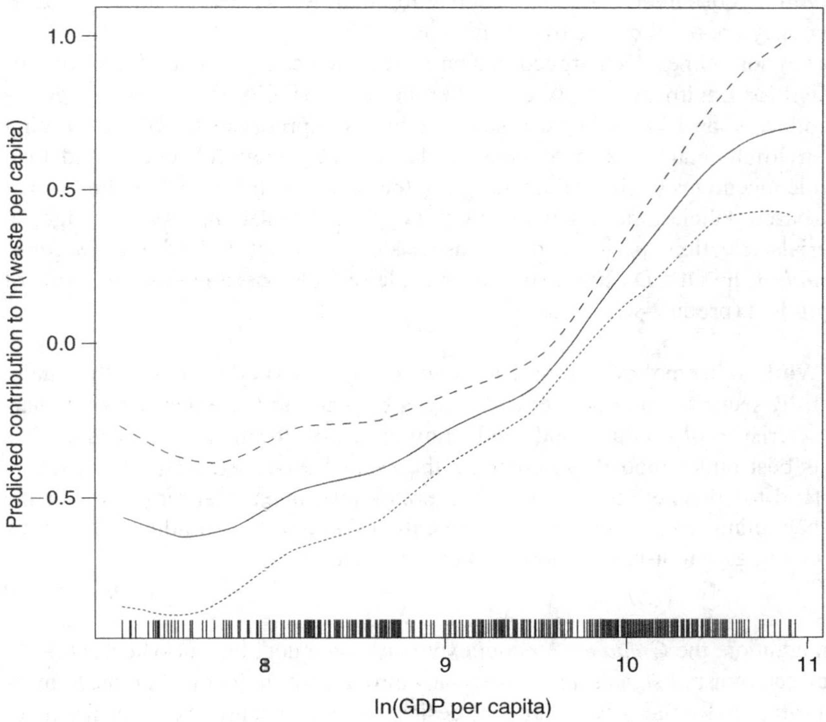

Figure 3.1 Municipal waste generation: EKC semi-parametric results.

Note
The solid curve represents the estimated line and the broken-line curves correspond to upper and lower 95 percent confidence intervals.

for a given country. We find the term statistically significant. Observing the positive relationship at middle and high incomes, we explicitly do not support the EKC hypothesis.

3.3 Why an EPR policy?

The primary advantage of an EPR policy over the DF policy is that it can eliminate an incentive for illegal disposal. Suppose that the DF policy is adopted, or that responsibility for recycling and the disposal of end-of-life goods is transferred from municipalities to consumers. Note that end-of-life goods are usually considered to be waste in developed countries; that is, their prices are negative. As a result, both consumers and waste management firms have an incentive to dump them (see Chapter 2). An EPR policy, however, requires that producers organize the collection, recycling, and waste disposal of end-of-life goods. Producers are also responsible for the recyclers. Thus the incentive for WMFs to dump waste diminishes in response to monitoring and control by the producers.

In addition, consumers have no incentive to dump waste, because under an EPR policy they are not required to pay any cost.

It has sometimes been argued that an EPR policy can efficiently improve DfE (Design for Environment). By extending the responsibility for end-of-life goods to producers, an EPR policy can send producers appropriate signals concerning the environmental impacts of their products during their life cycles and then provide incentives to change the design of the products. It is widely believed that EPR-based policies, more than the DF policy, can stimulate upstream changes in materials selection and in the design aspects of a product. Indeed, the *Guidance Manual* of the OECD (2001) stresses the role of EPR-based policies in terms of the product (product-waste) chain:

> While other policy instruments tend to target a single point in the chain, EPR seeks to integrate vertically signals related to the environmental characteristics of products and production processes throughout the chain. This is best understood if we compare the usual EPR-based instruments which tend to "bracket" the chain with a policy instrument that targets either the beginning (e.g., virgin material tax), the middle (e.g., a product tax), or the end (e.g., a unit-based waste fee) of the chain.
>
> (*Guidance Manual*, p. 48)

In addition, the *Guidance Manual* expresses some doubt about whether the DF policy can transmit signals all the way back through consumption patterns, to manufacturing, and ultimately to product design. "In some sense the enthusiasm for EPR is based on a belief that such signals are not transmitted effectively through arm's-length market transactions. Thus, there is a belief that a unit-based waste fee will not result in sufficient incentives for DfE" (*Guidance Manual*, p. 49).

Is this judgment true? Does the DF policy fail to improve DfE? We shall investigate this theoretically in the next chapter, but meanwhile, in the next two sections, we first of all briefly explain a few types of policies for implementing EPR, and then we give some examples of specific implementations of such policies.

3.4 EPR-based policies

There are several policies consistent with EPR. Our summary descriptions below rely on explanations given in the *Guidance Manual.*

Product take-back requirement

The most active application of the concept of EPR, under both voluntary and mandatory schemes, is a policy that obliges producers to take back their products. The concept may be applied to specific products (e.g., automobiles, electric appliances, electronic products) or to waste (e.g., automobile shredder residue (ASR)), all of which are to be taken back. A take-back requirement scheme often entails the setting of targets for collection and recycling. Usually responsibility for meeting the reuse, recycling, and collection targets is laid upon the producers of those goods, unless other conditions—such as participation in a producer responsibility organization (PRO)—have been met (*Guidance Manual*, p. 40).

While a take-back requirement directly imposes responsibility for recycling or disposing of end-of-life goods on the producer, certain economic instruments can also be used toward the same ends. These include deposit-refund schemes, advance disposal fees, and taxes and/or subsidies (*Guidance Manual*, p. 41).

Deposit-refund schemes

Under a deposit-refund system in the strict sense, an extra charge is paid when the product is purchased, and this charge is refunded when that product is returned to the retailer or to a specified treatment facility. Although deposit-refund schemes have long been adopted mostly with a view to recycling bottles and cans in which beverages are sold in the retail market, they have also been adopted in regard to such products as batteries, fluorescent lights, automobile tires, and shopping bags (*Guidance Manual*, p. 42).

Upstream combination tax/subsidy

At the 1998 EPR Workshop in Washington, DC, an "upstream combination tax/subsidy" (UCTS) model was suggested as a possible alternative economic instrument for promoting EPR. This model would entail the imposition of a tax on intermediate products (for example, on aluminum ingots or on rolls of paper of a specific grade) along with the payment of cash to those who collected and brought to a specified treatment facility end products containing such intermediate products (for example, aluminum cans or old newspapers). This kind of a tax/subsidy scheme provides an incentive to producers to change the materials they use and their product design, and at the same time provides a mechanism for funding (and thus encouraging) recycling and treatment (*Guidance Manual*, p. 44).

Advance disposal fee

Though at first glance it seems a contradiction, an advance disposal fee (ADF), which consumers must pay at the time they purchase certain products, can be used as an instrument to promote EPR if producers are required by law to recycle their products. Producers (and importers) can, through a private sector body, ensure that their products are collected and dealt with properly. The fees earlier paid by consumers are used for subsidizing these recycling and disposal activities. An ADF can therefore also be considered a tax/subsidy combination (*Guidance Manual*, pp. 42–43).

Because both the UCTS and the ADF are combinations of a tax and a subsidy, they can, in a way, be considered deposit/refund systems. In the following chapters, therefore, we shall understand the UCTS and the ADF to be deposit/refund systems in the broad sense of the term.

As well as economic instruments such as the above, there are several standards that are considered to be consistent with EPR. One of these standards is a minimum recycled content requirement.

Minimum recycled content requirements

Manufacturers can be set a target of including in their products a certain minimum amount of recycled material. They could also be encouraged to retrieve their products so that the materials can be recycled again. This performance standard encourages manufacturers to investigate ways to effect changes in product design. A performance standard of this nature is used among manufacturers of paper products, glass containers, and plastic drink containers (*Guidance Manual*, p. 44).

3.5 Examples of how disposal fee and EPR-based policies are implemented

Western Canada used oil program[1]

To deal with the problem of improper disposal of used oil in the western provinces of Canada, a program was established in 1997 by which those in the oil industry—the sellers and importers of oil, oil filters, and oil containers—would be required to pay an "environmental handling charge" (EHC). The aim was to encourage the recycling of the oil, filters, and containers as much as possible, since oil can be refined or simply reused in things like small space heaters, while filters can be recycled into steel and plastic oil containers can be turned into plastic goods. Authorized collectors and recyclers of used oil, filters, and containers were paid what was called a "return incentive."

Collection of the EHC and payment of the return incentives are carried out by nonprofit organizations. The amounts of the EHC and the incentives vary from province to province; the cost of transporting the used items is an important factor because of distances involved. The program can be considered a deposit/refund scheme.

Since the introduction of the program, recovery rates have risen every year. There is also some evidence that the program has promoted DfE: some oil companies have changed the content of their products (for example, by removing lead from lubricants) or have switched to thin-wall plastic containers.

The UK packaging waste program[2]

A program that is working well in the United Kingdom is a tradable credit scheme applied to packaging waste. In 1997 the government passed Producer Responsibility Obligations (Packaging Waste) Regulations and in 1998 Packaging (Essential Requirements) Regulations; these oblige producers to recycle a specified percentage of their packaging waste each year, and that percentage has been steadily raised since then. In 2004, for example, the recovery target was set at 18 percent for wood, 21.5 percent for plastics, and 65 percent for paper. The regulations divide producer responsibility into four categories—manufacturer, converter, packer/filler, and seller—and each category is assigned its own recycling obligations.

A feature of the program is that manufacturers, converters, packers/fillers, and sellers can participate in a "compliance scheme" that will fulfill the obligations for them. Most companies join a compliance scheme rather than act individually. Compliance schemes (as well as individual companies) receive from recyclers packaging waste recovery notes (PRNs) that vary according to what is recycled (there are PRNs for glass, paper, aluminum, steel, plastics, wood, general recycling, and energy recovery). Generated every time a ton of packaging material is recycled, the notes prove that compliance schemes/individual companies are complying with the regulations. PRNs can be traded among obligated individual companies, compliance schemes, and recyclers, within an organized market called the Environmental Exchange. Accredited recyclers must report quarterly to the government how many tons of packaging they reprocessed in the quarter; the government then issues blank notes to these recyclers, and they can issue PRNs only up to the amount of recycling that they have reported to the government. (Packaging waste that is exported for recycling is acknowledged by means of a packaging export recovery note, or PERN. PERNs are also considered acceptable evidence of compliance.)

Producers that use more resources in their product packaging must hold more PRNs or PERNs. The price of a PRN or PERN is considered a tax on packaging. Because recyclers issuing the notes receive the price, it is considered a subsidy for recycling packaging waste. The UK Packaging Waste Program can, therefore, also be regarded as a deposit/refund system, in which both the deposit and the refund are equivalent to the price of a PRN or PERN.

Recycling and recovery have increased since the regulations came into effect. The increase in recycling rates ranges between 45 percent and 137 percent. The program does provide incentives for downsizing and some substitution of eco-friendly materials. Producers find it profitable to choose materials that are easier to recycle, because the PRN price then becomes lower.

The Dutch Waste Electrical and Electronic Equipment (WEEE) Program[3]

An EPR policy for end-of-life electrical and electronic equipment was introduced in the Netherlands in 1998. The program is based on regulations regarding white and brown goods that obliged retailers to take back end-of-life electrical and electronic goods in exchange for new ones and that obliged producers to accept those products from retailers and recycle them. Recovery and reuse targets ranged from a high of 75 percent for refrigerators down to 45 percent for small appliances.

The program adopts an ADF system in order to cover recycling costs. In the case of household appliances, stereos, and television sets, the producer responsibility organization (PRO) that handles collection and recycling for most producers has charged fees ranging from zero for small household products up to $17 for refrigerators and some television sets. Computer equipment has been handled by a separate PRO, which charges a fee on the basis of weight.

Recycling rates have been high; for example, the rate for refrigerators was 86 percent in 2000 and 85 percent in 2001, compared with a target of 75 percent, and that for small appliances ranged from 60 to 64 percent in both years, compared with a target of only 45 percent.

Home appliances recycling program in Japan[4]

The current recycling system governing home appliances in Japan was established through the Law for Recycling of Specific Household Appliances (Household Appliances Recycling Law). The law, enforced in 2001, covers four items: air conditioners, refrigerators, television sets, and washing machines. The law obliges both producers and retailers to collect, transport, and recycle these household appliances.

In those cases in which retailers are asked to collect the appliances, retailers must collect the goods and pass them on to the producers that manufactured them. When producers are asked to collect the appliances that they produced, they must accept and recycle them. The target recycling rates are 70 percent for air conditioners, 50 percent for television sets, and 65 percent for washing machines.

In contrast to the EPR-based policies that have been adopted in many OECD countries, Japanese consumers (disposers) of items must pay fees for both collection and recycling at the disposal stage. The law is based on the DF policy. The fees were set for each category in 2001. Recycling fees were set at ¥3,500 ($35) for air conditioners, ¥4,600 ($46) for refrigerators, ¥2,700 ($27) for television sets, and ¥2,400 ($24) for washing machines, while collection fees were set at ¥1,500–2,500 ($15–25) for air conditioners, ¥1,500–3,000 ($15–30) for refrigerators, ¥1,000–2,000 ($10–20) for television sets, and ¥1,500–2,500 ($15–25) for washing machines.

Hosoda (2004) points out that several product design changes eventuated as a result of the law's implementation. The first change in design was for easier dismantling. The connection and fixing methods used in household appliances have been improved so that the appliances can be dismantled more easily and speedily. Plastic bolts and other connecting devices have replaced metal ones in predominantly plastic appliances to reduce sorting times. Many different kinds of plastics were once used in household appliances, making it time-consuming to sort parts by type of plastic, so manufacturers have been reducing the number of different materials in their products since the law came into effect.

The second change in product design has been a "shorter flow back into the recycling stream." Before the law was enacted, materials recovered from household appliances were used as lower-grade materials. But now the manufacturers try to recycle the recovered materials directly into the same kind of products/parts. This change makes the flow of waste back into the resources stream shorter and saves on energy consumption. Examples of such direct reuse are glass from cathode ray tubes (CRTs) put back into CRTs and tanks from washing machines put back into tanks for new washing machines. Other changes of product design include the development of non-CFC refrigerators and lead-free soldering.

The list of fees for recycling was updated in 2009.[5] According to it, recycling fees have substantially decreased. The new recycling fees are ¥2,500 ($25) for air conditioners, ¥3,600–4,600 ($36–46) for refrigerators, ¥1,700–2,700 ($17–27) for television sets, and ¥2,400 ($24) for washing machines. Part of the reduction stems from changes that have been made in product design.

It is estimated that in Japan 22.87 million end-of-life household appliances were generated in 2005, and only around half (10.76 million) were recycled domestically. A large proportion of the remainder (7.35 million) was exported as used goods to developing countries, and a small portion (0.1 million) was illegally disposed of.[6] There are huge used-goods markets in developing countries, where many end-of-life, but still usable, appliances from developed countries are sold. The international trade in used goods is a complicated issue and we deal with it later as a separate topic.

End-of-life vehicles recycling program in Japan[7]

In Japan about five million end-of-life vehicles are disposed of each year. Since they contain useful metals and parts, they are usually recycled or reused. End-of-life vehicles used to be bought by companies that dismantled them and resold usable parts as used parts, while the rest of the end-of-life vehicles were sold to shredding companies, which broke them down further with shredding machines and roughly sorted the materials for recycling.

Until 1990 the price of end-of-life vehicles was positive. However, it became widely recognized that shredder dust contains heavy metals, so it must be properly disposed of. It was also recognized that the CFC used in air conditioners in vehicles must be dealt with in such a way that CFC does not get into the air. Since 1996 shredder dust must be disposed of at well-managed disposal sites, and since 2002 the CFC must be recovered. These regulations sharply pushed up dismantling costs and the final disposal costs of automobile shredder residue (ASR), and on top of this, the price of scrap iron had, until 2001, been falling. These trends in scrap price and disposal cost meant significantly lower profit for shredding companies. In the end, the value of end-of-life vehicles fell to such an extent that recycling them produced no profits at all; they were of negative value. As a result, in the 1990s the illegal dumping of end-of-life vehicles became rampant in Japan. They were being illegally dumped by end users, dismantling companies, and shredding companies.

To address this problem the government of Japan enacted the Law for the Recycling of End-of-Life Vehicles, which came into effect in 2005. The law adopted an advance disposal fee (ADF) system. Purchasers of new vehicles were now required to pay a disposal fee in advance. The fee covers mainly three items: the cost of disposing of ASR, the cost of air bags, and the cost of recovering the CFC (including an information management cost and a finance management cost). Although the total fee varies with the brand, size, type of engine (normal or diesel), and year of production, it comes to about ¥10,000 for domestic vehicles and ¥20,000 for imported vehicles.

The one feature of the law is that it applies not to the vehicles but to the waste, i.e., ASR and CFC. The law requires that ASR be recycled at 30 percent by 2005, 50 percent by 2010, and 70 percent by 2015. Another important feature of the law is that if end users export end-of-life vehicles overseas, the recycling fee that they paid in advance will be returned to them.

3.6 Responses of householders to a disposal fee

As mentioned in Section 3.1, we can also reduce waste emission by means of a disposal fee. However, though the *true* waste emission may be reduced only moderately, the *legal* waste disposal might be reduced drastically. The difference in the two reductions might be illegal disposal. In this section we shall examine empirically how households respond to a disposal fee.

Elasticities of waste emission and the possibility of illegal disposal

Many empirical studies investigate how much a disposal fee can reduce household or municipal waste. They estimated elasticities of (legal) waste emission with respect to the own price, income, and cross-price. The own price elasticity indicates the percentage of waste reduction in response to increasing the price of waste collection by 1 percent (measured by disposal fee). Choe and Fraser (1998) reviewed the literature, and the estimation results in most empirical studies indicate that households reduce waste emission after the introduction of a disposal fee. But the estimated own price elasticity is not as high as expected (from –0.6 to –0.058; see Table 1 in Choe and Fraser, 1998).

Households have various options in response to a disposal fee: reducing consumption of goods that yield larger amounts of post-consumption waste (or demanding less packaging); more recycling; more composting; throwing waste into commercial dumpsters (large trash receptacles); illegal disposal; and illegal burning. The first three options are legal, the last three are not.

How do households respond to the introduction of a disposal fee? Fullerton and Kinnaman (1996) investigate what happened when a disposal fee was introduced in December 1991 in Charlottesville, Virginia. They found that in response to the disposal fee program the average household reduced the weight of waste by 14 percent, reduced the volume of waste by 37 percent, and increased the weight of recycling by 16 percent. In addition, they found indirect evidence that illegal dumping may have accounted for 28–43 percent of the reduction in waste. They concluded that a disposal fee has little effect on the weight of waste but it has substantial effects on volume, density, recycling, and illegal dumping. Households would respond to a disposal fee by compacting waste into fewer bags, with the result that the volume and density of waste in each bag would increase. This phenomenon is known as the "Seattle Stomp," and it depends on whether the disposal fee is based on weight or on volume.

Usually a disposal fee is levied on the size of a waste can or bag, which means that the disposal fee is not based on the weight of the waste but on its volume.

The experience in Japan

In Japan the number of municipalities adopting a disposal fee on household waste has increased. In 2006 the number was 1,840 (52.9 percent of all municipalities). In Japan the definition of waste is based on the Waste Disposal and Public Cleansing Law (1970), which defines industrial waste as waste emitted in economic activities such as production and construction, and other waste is defined as municipal waste. The latter refers to household waste, ordinary business waste, and recyclable waste. Ordinary business waste is what is emitted by small-scale businesses, and it includes waste from shops, supermarkets, restaurants, schools, and service areas along highways.

We investigated data from 48 municipalities that implemented a disposal fee program from 1999 to 2004. In these programs a disposal fee was levied on household waste, but the level of the fee varied across municipalities. Our data covers the amount of municipal waste (household waste, ordinary business waste, and recyclable waste). We calculated each kind of waste emitted (per capita and per day) both in the year prior to the introduction of the disposal fee

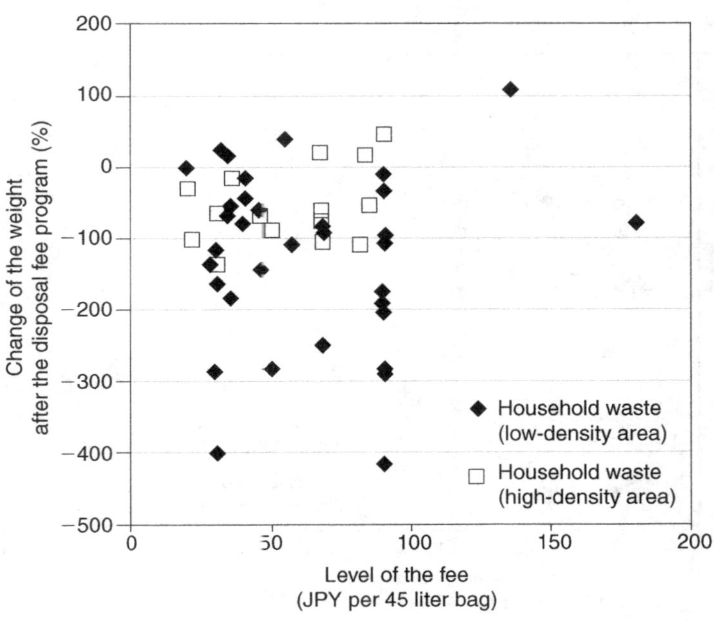

Figure 3.2 Change in the weight of household waste (per capita and per day).

Note
1US$ = 100¥.

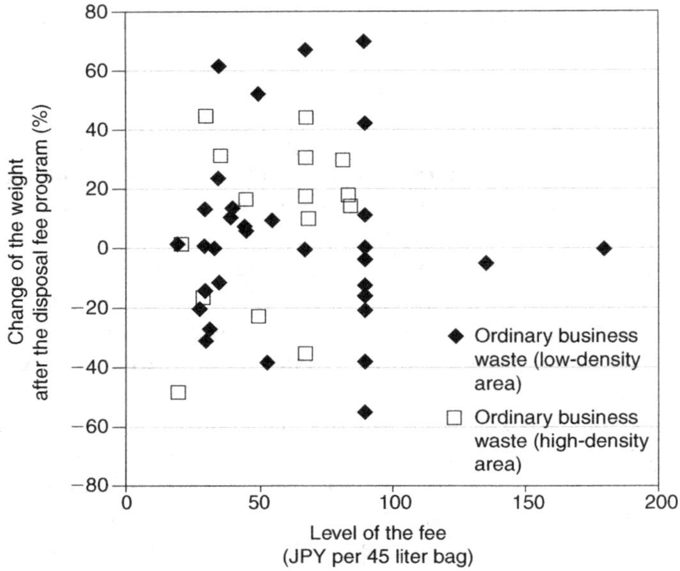

Figure 3.3 Change in the weight of ordinary business waste (per capita and per day).

Note
1US$ = 100¥.

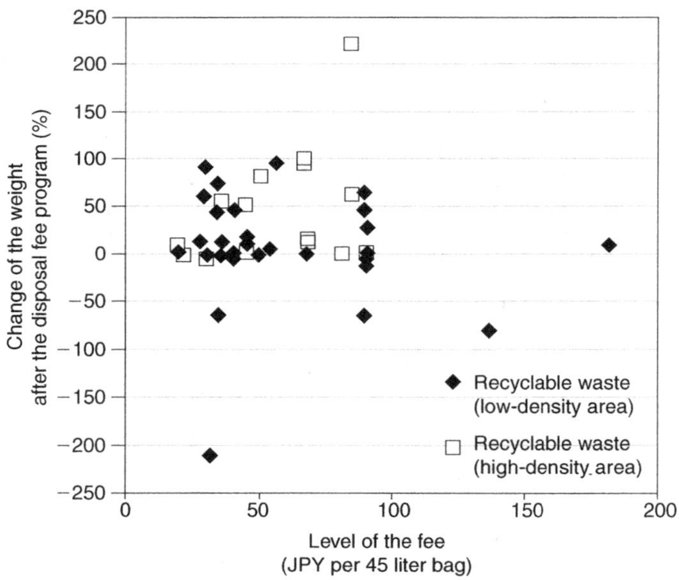

Figure 3.4 Change in the weight of recyclable waste (per capita and per day).

Note
1US$ = 100¥.

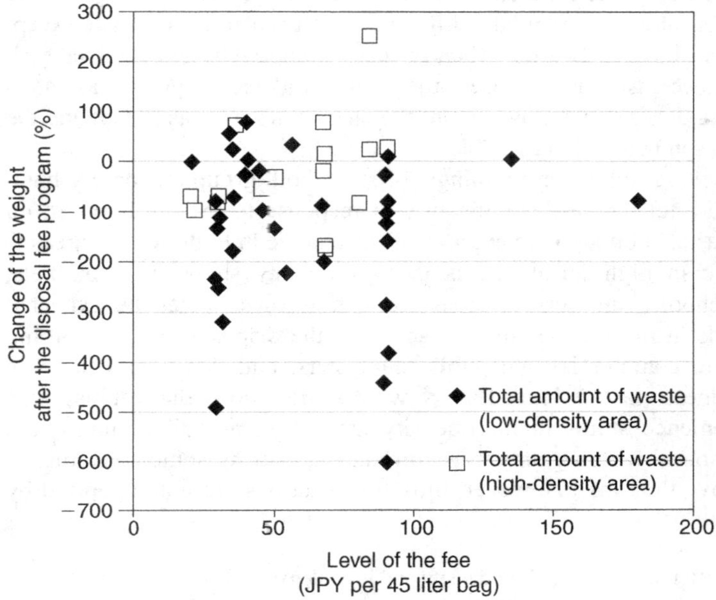

Figure 3.5 Change in the weight of total amount of waste (per capita and per day).

Note
1US$ = 100¥.

Table 3.3 Descriptive statistics for each type of waste and each group

Variables	*Group*	*Obs*	*Average*	*Std. error*
Total amount of waste	Low-density	33	−133.74 g	165.80
	High-density	15	−1.34 g	90.86
Household waste	Low-density	33	−125.97 g	123.76
	High-density	15	–53.67 g	53.36
Ordinary business waste	Low-density	33	−10.09 g	128.55
	High-density	15	6.80 g	28.66
Household + ordinary business waste	Low-density	33	−136.06 g	163.37
	High-density	15	–46.86 g	47.65
Recyclable waste	Low-density	33	2.32 g	53.43
	High-density	15	45.52 g	61.18

and in the year after its introduction, and then calculated the difference between the two years. We divided the 48 municipalities into two groups, the first being areas of low population density (fewer than 2,000 people per square kilometer) and the other being areas of high population density (more than 2,000 people per square kilometer). The data is plotted in Figures 3.2 to 3.5 and the descriptive statistics are given in Table 3.3. Unfortunately, there is no statistically significant relation between the changes in the volume of waste and the levels of the fee, even when the data for the two groups is looked at separately.

Next, we examined the difference between the two groups. As shown in the last column of Table 3.4, we found a difference in all categories of waste except that of "ordinary business waste." Then for each group we tested whether each type of waste decreased or increased after a disposal fee program was introduced. The hypothesis that the average of the change is zero was tested, and the results are also summarized in the table.

There are several interesting findings. First, in both groups "ordinary business waste" did not increase significantly in response to the fee imposed on "household waste," although the average of the change in high-density areas is positive. Since in high-density areas there are many shops, supermarkets, restaurants, schools, and service areas along highways, it can be expected that households in those areas might respond to the disposal fee by dumping their waste into commercial and public dumpsters, and this would increase "ordinary business waste." Indeed, there were some reports that rubbish bins outside convenience stores in high-density areas became full of household waste after disposal fee programs were introduced, causing some convenience stores to remove the bins.[8] However, those expectations are not supported by the data.

Table 3.4 The difference between low-density and high-density areas

	Low-density	*High-density*	*Difference between the two groups*
Total amount of waste	Decreasing t = –4.634, p-value = 5.74e-05	No change t = –0.0573, p-value = 0.9551	Yes t = –3.560, p-value = 0.0009 (Welch Two Sample t-test)
Household waste	Decreasing t = –5.847, p-value = 1.69e-06	Decreasing t = –3.895, p-value = 0.00162	Yes t = –2.827, p-value = 0.00693 (Welch Two Sample t-test)
Ordinary business waste	No change t = –0.451, p-value = 0.6551	No change t = 0.9188, p-value = 0.3738	No t = –0.717, p-value = 0.478 (Welch Two Sample t-test)
Household + ordinary business waste	Decreasing t = –4.784, p-value = 3.72e-05	Decreasing t = –3.809, p-value = 0.001917	Yes t = –2.879, p-value = 0.006275 (Welch Two Sample t-test)
Recyclable waste	No change t = 0.249, p-value = 0.8046	Increasing t = 2.882, p-value = 0.01206	Yes t = –2.482, p-value = 0.0168 (Two Sample t-test)

Second, "recyclable waste" increased (or recycling was promoted) in high-density areas, while it did not increase in low-density areas. This means that it is inefficient to collect "recyclable waste" in low-density areas.

Third, "household waste" decreased dramatically in low-density areas, which reduced the "total amount of waste" in the same areas. On the other hand, while "household waste" also decreased significantly in high-density areas, the "total amount of waste" remained constant. In those areas the increase in recycling offset the decrease in the "household waste."

Because the "total amount of waste" to a certain extent reflects the standard of living, it is expected to remain constant even after the introduction of a disposal fee program. Curiously, however, the "total amount of waste" significantly decreased in low-density areas, while it remained constant in high-density areas. Noting that illegal disposal and illegal burning are two methods that can easily be resorted to in low-density areas, we can expect some increase in illegal dumping or illegal burning in response to a disposal fee in low-density areas. And as a matter of fact there have been some reports of an increase in illegal disposals in those areas after the introduction of a disposal fee.[9]

According to Table 3.3, "household waste (per capita and per day)" decreased on average by 133.74 g in low-density areas. We cannot conclude, however, that all of the decrease is attributable to illegal activities such as dumping or the burning of plastic waste. Some of the decrease can be attributed to the disposal of raw kitchen garbage in backyards for composting, because residents in low-density areas usually have comparatively large yards or gardens.

4 First-best economy

Comparisons between a disposal fee policy and EPR-based policies

4.1 Introduction

In this chapter we compare three policies that might be used in the first-best economy, that is, one in which there is no possibility that consumers will illegally dispose of end-of-life-goods in response to a disposal fee. The three policies considered in this chapter are a disposal fee (DF) policy, a deposit/refund (DR) system, and a product take-back (TB) requirement. The latter two policies are based on extended producer responsibility (EPR). We show that the three policies are equally effective in the first-best economy. This conclusion implies that EPR is superior to DF, because in every actual economy illegal disposal *is* possible.

There are two key works closely related to the topics; one is that by Fullerton and Kinnaman (1995) and the other is that by Fullerton and Wu (1998). The former work was the first published paper to compare the three policies theoretically. We summarize their results in Section 4.3. The latter work extends the former by incorporating Design for Environment (DfE) into the model. We review the theoretical results in Section 4.4.

While the theories introduced in this chapter are based on general equilibrium models, we can grasp the main conclusion intuitively in the partial equilibrium model that is developed in the next section.

4.2 A basic result in a partial equilibrium model

Under a simple setting we can see that DF and EPR are equivalent to each other and that each can achieve the social optimum. To see this, we need to compare in a simple partial equilibrium model four policy options with one another: 1) provision by the municipality of a waste management service as a public good; 2) a DF policy; 3) a product take-back (TB) requirement; and 4) a deposit/refund (DR) system.[1] EPR policies are represented by the last two options. Note that "deposit/refund system" is used in its broader sense, to include an upstream combination tax/subsidy (UCTS) and an advance disposal fee (ADF).

We sketch a product market in Figure 4.1. Denote by C_d the waste management cost per unit of end-of-life goods, which is represented by the length of E_2E_3 in the figure. In Case 1, neither producers nor consumers are required

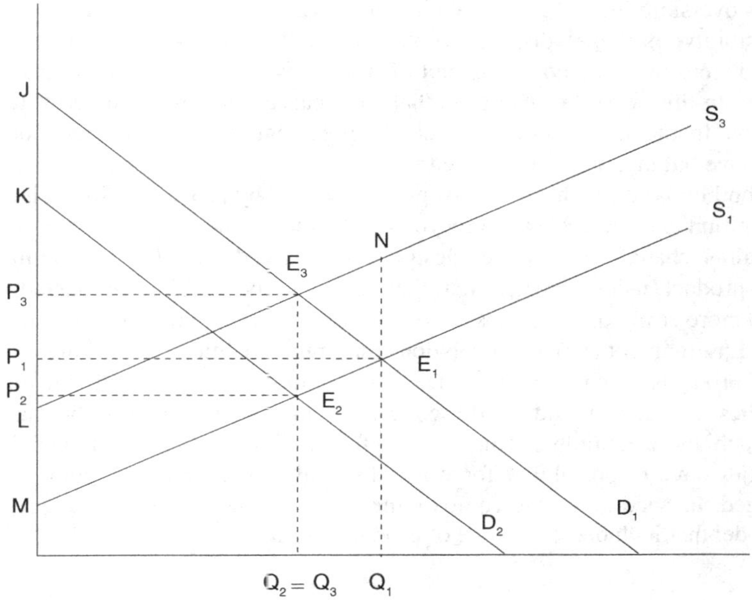

Figure 4.1 Market equilibrium under four policy options: free waste collection by the municipality, a DF policy, a TB requirement system, and a DR system.

to pay the waste management cost, so the equilibrium is E_1. The sum of the producer surplus and the consumer surplus is ΔJE_1M. The waste management cost incurred by the municipality is $\square LME_1N$. Therefore, the social welfare is ΔJE_3L minus ΔE_3E_1N.

In Case 2, the demand curve shifts downward from D_1 to D_2, because consumers take into account the disposer fee (E_2E_3) in advance when they purchase goods. Note that the equilibrium is E_2. Then the social welfare is represented by ΔKE_2M. Note that the consumer surplus is already reduced by the waste management cost represented by $\square JKE_2E_3$. Because ΔKE_2M is equal to ΔJE_3L, it follows that the social welfare can be increased by ΔE_3E_1N when the municipality quits providing the waste management service as a public good.

In Case 3, consumers need not pay for the waste management cost, but the producers pay for it. The requirement pushes the supply curve upward to S_3. The new equilibrium is E_3. The social welfare in this case is ΔJE_3L, which is equal to that of Case 2.

In Case 4, consumers are required to pay the waste management cost as a tax when they purchase goods and then the cost is provided to waste management firms as a subsidy. Therefore, the demand curve shifts downward from D_1 to D_2, which moves the equilibrium to E_2. Obviously, all the results are the same as in Case 2.

Thus Case 1 is inferior to the other cases. In addition, it can easily be shown that Case 2, Case 3, and Case 4 are socially optimal in the first-best economy where there is no possibility of illegal disposal. The model analyzed above, however, is oversimplified. At least two important aspects are omitted. The first is the possibility of illegal disposal. Note that in the model it is implicitly assumed that consumers do not dump end-of-life goods illegally in response to a disposal fee. In this sense, we can state that EPR-based policies are superior to a DF policy. In the next section, we shall implement the same comparison analyzed above but in a more general setting.

The second omission is the choice of product design by producers. Reducing waste generation and increasing the recycling rate of end-of-life goods to some extent requires changes to product design, such as reducing the packaging around the product, using recycled materials as input, and enabling products to be recycled more easily or at less cost. These aspects of product design are called Design for Environment (DfE). It is obvious that municipalities cannot improve the DfE of products. What is more, it has become more and more difficult for municipalities to recycle end-of-life goods, because their composition has become much more complex. This is why the traditional system, in which municipalities have responsibility for waste disposal and recycling, is going to be challenged. In Section 4.4 we compare the three policies in a more comprehensive model that includes the choice of product design.

4.3 Basic general equilibrium model 1 (Fullerton and Kinnaman, 1995)

The setting of the model

Let's assume a competitive economy with n homogeneous consumers. The utility of a representative consumer depends on: consumption of goods denoted by c; leisure (time) denoted by k_h; and total amount of garbage, $G = ng$.

$$U = U(c, k_h, G). \tag{4.1}$$

It is assumed that $U_c > 0$, $U_h > 0$, and $U_G < 0$. The consumer can either sell end-of-life goods to producers for recycling or use a waste management service provided by waste management firms for disposal. Therefore, the mass balance condition must hold:

$$c = g + r. \tag{4.2}$$

End-of-life goods to be recycled and those to be disposed of are represented by r and g respectively. Contrary to Fullerton and Kinnaman (1995), we assume that preparation is needed for recycling; this includes separation, storage, and transportation of end-of-life goods. It is assumed that m units of labor are needed for the preparation of one unit of end-of-life goods to be recycled. Let k_r stand for labor inputs at the preparation stage. Then the following equation must hold:

$$r = mk. \tag{4.3}$$

The consumption goods are produced by competitive producers using labor k_c and recycled materials r. The production function of goods is represented by $c = f(k_c, r)$, which has constant returns to scale.

The waste management services are provided using only labor denoted by k_g. The technology is assumed to have constant returns to scale and it is assumed that one unit of labor collects γ unit of waste, i.e.,

$$g = \gamma k_g. \tag{4.4}$$

The labor resource constraint must hold:

$$k = k_c + k_g + k_r + k_h, \tag{4.5}$$

where k denotes a fixed total labor.

Social optimum

By substituting (4.3), (4.4), and (4.5) into equation (4.1) and (4.2), the Lagrangian for the social optimal problem can be written as

$$L = U(\gamma k_g + mk_r, k_h, n\gamma k_g) + \delta\{f(k - k_g - k_r - k_h, mk_r) - \gamma k_g - mk_r\},$$

which is maximized with respect to k_g, k_r, and k_h.

The first-order conditions (FOCs) are

$$U_c + U_G n = \delta(1 + f_k / \gamma) \tag{4.6}$$

$$U_c = \delta(1 - f_r + f_k / m) \tag{4.7}$$

$$U_h = \delta f_k. \tag{4.8}$$

Market equilibrium under three policies

Because there is a negative externality associated with the total waste G, the competitive market equilibrium is distorted away from the social optimum. To restore the social optimum in the decentralized market, the government has at its disposal three policy instruments: tax levied on consumption (t_c); disposal of waste (t_g); and recycling of waste (t_r). These taxes can be negative (i.e., a subsidy). We consider three types of policy sets: a) a DF policy with $t_g>0$; b) a DR policy with $t_c>0$ and $t_r<0$; and c) a TB policy.

DF policy

Under this policy, consumers choose g, r, and k_h to maximize utility,

$$U(g + r, k_h, G),$$

subject to the budget constraint

$$p_k(k - r / m - k_h) = (1 + t_c)(g + r) + (p_g + t_g)g - (p_r - t_r)r.$$

The wage rate is denoted by p_k. End-of-life goods can be either disposed of or recycled. If the consumer disposes of one unit of end-of-life good, the waste management fee (p_g) must be paid in addition to the disposal tax (t_g). Alternatively, if end-of-life goods are sold to a producer for recycling, the consumer can receive the price (p_r). Note that consumption goods are the numeraire (the price is 1).

The FOCs for the consumer's problem are

$$U_c = \lambda(1 + t_c + p_g + t_g) \tag{4.9}$$

$$U_c = \lambda(1 + t_c - p_r + t_r + p_k / m) \tag{4.10}$$

$$U_h = \lambda p_k, \tag{4.11}$$

where the Lagrange multiplier is denoted by λ.

Producers choose k_c and r to maximize profits:

$$\Pi_c = f(k_c, r) - p_r r - p_k k_c.$$

The FOCs are

$$f_k - p_k = 0 \tag{4.12}$$

$$f_r - p_r = 0. \tag{4.13}$$

WMFs maximize $\Pi_g = p_g g - p_k k_g = (p_g \gamma - p_k)k_g$. Then we obtain

$$p_g = p_k / \gamma. \tag{4.14}$$

By a simple arrangement of these FOCs, we can obtain

$$U_c = \lambda(1 + t_c + f_k / \gamma + t_g) \tag{4.15}$$

$$U_c = \lambda(1 + t_c - f_r + t_r + f_k / m) \tag{4.16}$$

$$U_h = \lambda f_k. \tag{4.17}$$

To attain the social optimum in the decentralized market, the competitive market equilibrium conditions (4.15) to (4.17) must be coincident with (4.6) to (4.8), respectively. We can easily see that the optimal DF policy is characterized as

$$t_g = -nU_G / \lambda > 0 \text{ and } t_c = t_r = 0. \tag{4.18}$$

Note that (4.8) and (4.17) indicate that $\delta = \lambda$. Thus, when there is no possibility of illegal disposal, only a disposal tax is necessary and no other policy instruments are needed.

DR policy

As long as $p_g + t_g > 0$, consumers have a strong incentive to dispose of waste illegally. Fullerton and Kinnaman (1995) raise an interesting question: whether the government can attain the social optimum when free disposal services are provided, or $p_g + t_g = 0$. They show that the answer is yes. Indeed, the social optimum can be attained if the government sets t_c and t_r as follows:

$$t_c = (f_k - nU_G) / \gamma > 0 \tag{4.19}$$

$$t_r = -t_c < 0. \tag{4.20}$$

This policy set is a deposit/refund system. Note that if t_G and t_r are set as (4.19) and (4.20), the competitive equilibrium conditions (4.15) to (4.17) coincide with the social optimum conditions (4.6) to (4.8). Consumers pay a deposit (t_c) at the purchase stage of goods and they will be paid a refund ($-t_r > 0$) when they sell the end-of-life goods to producers.

TB policy

Adopting a product take-back policy is another solution for discouraging illegal disposal. Under this policy, the problems of producers and a representative consumer are revised as follows.

Producer's problem:

$$\text{Max } \Pi_c = f(k_c, r) - p_k k_c - (p_g + t_g)(f(k_c, r) - r) - t_r r - p_k r / m$$

with respect to k_c and r.
The FOCs are

$$f_k - p_k - (p_g + t_g) f_k = 0 \tag{4.21}$$

$$f_r - (p_g + t_g)(f_r - 1) - t_r - p_k / m = 0. \tag{4.22}$$

Consumer's problem:

$$\text{Max } U(c, k_h, g),$$

subject to $p_k k = (1 + t_c) c + p_k k_h$.
The FOCs are

$$U_c = \lambda(1 + t_c) \tag{4.23}$$

$$U_h = \lambda p_k = \lambda\{1 - (p_g + t_g)\} f_k. \tag{4.24}$$

Note that the last equality is obtained from (4.21). The profit maximization problem for the waste management firm remains the same and so equation (4.14), $p_g = p_k / \gamma$, is still valid.

The optimal policy set is characterized by

$$t_c = f_k / \gamma - nU_G / \lambda > 0 \tag{4.25}$$

$$t_g = -f_k / \gamma < 0 \tag{4.26}$$

$$t_r = -t_c < 0. \tag{4.27}$$

It can be easily checked that (4.23) coincides with (4.6). To see that (4.8) is satisfied, it must be that $p_g = -t_g$. We can easily check this, because substituting (4.14) and (4.26) into (4.21) yields $p_g = f_k / \gamma = -t_g$. Finally, for (4.7) to be satisfied, it is sufficient that $1 + f_k / \gamma - nU_G / \delta = 1 + f_k / m - f_r$. Note that, if it is satisfied, (4.23) coincides with (4.7). We can easily see that this equation coincides with (4.22) if we set $t_r = -t_c$ as stated in (4.27).

Note that the disposal tax (t_g) and the tax on recycling (t_r) are not levied on the consumer but on producers. Under a TB policy, producers are required to receive end-of-life goods from the consumer at no charge. What is more important, a TB policy alone cannot attain the social optimum and it must be linked with other policy instruments, with subsidies for (legal) disposal and recycling that will be paid to producers.

4.4 Basic general equilibrium model 2 (Fullerton and Wu, 1998)

In this section we review Fullerton and Wu (1998) in order to investigate whether a DF policy and EPR-based policies (a DR policy or a TB policy) can promote any DfE of products. As in the preceding section, we once again consider a competitive economy with n homogeneous consumers. The consumption goods are also produced by competitive producers and disposal services are provided by WMFs. Both of their technologies are assumed to be constant returns to scale.

The DfE of products is represented by their recyclability, ρ, which is defined as the percentage of the product that can be recycled. The consumption goods are produced using labor k_c and recycled materials r. In its production decision, the producer also chooses the product's recyclability ρ. The production function of goods is thus represented by

$$c = f(k_c, r, \rho). \tag{4.28}$$

It is assumed that $f_k > 0, f_r > 0$, and $f_\rho < 0$. The last inequality implies that increasing the recyclability of the product needs some additional costs.

The utility of a representative consumer depends on consumption of goods denoted by c, leisure (time) denoted by k_h, and total amount of garbage, $G=ng$.

$$U = U(c, k_h, G)$$

It is assumed that $U_c > 0$, $U_h > 0$, and $U_G < 0$. It is assumed that the product recyclability does not directly affect the utility but indirectly through the resource constraint.[2]

A part of end-of-life goods ρc can be recycled and the other $(1-\rho)\ c$ cannot be recycled. Then we have two equations

$$r = \rho c \tag{4.29}$$

$$g = (1-\rho)\ c. \tag{4.30}$$

Note that consumers cannot choose the amount of end-of-life goods to be recycled, whereas they could in the model described in the preceding section. In addition, for simplicity, it is assumed that there is no need to prepare end-of-life goods for recycling.

The waste management services are provided using only labor. It is assumed that one unit of labor collects γ unit of waste,

$$g = \gamma k_g. \tag{4.31}$$

The labor resource constraint must hold:

$$k = k_c + k_g + k_h, \tag{4.32}$$

where k denotes a fixed total labor.

Social optimum

The object of the government is to maximize the utility of a representative consumer subject to (4.28), (4.29), (4.30), (4.31), and (4.32). The Lagrangian for the social optimal problem can be written as

$$L = U\,(c, k_h, ng) + \delta_1\,(f\,(k_c, r, \rho) - c) + \delta_2\,(r - \rho c) + \delta_3\,(g - (1-\rho)\ c)$$
$$+ \delta_4\,(\gamma k_g - g) + \delta_5\,(k - k_c - k_h - k_g).$$

Simple arrangements of the FOCs yield

$$-\frac{f_\rho}{f_k} - \frac{f_r}{f_k}c - \frac{1}{\gamma}c + \frac{nU_G}{U_h}c = 0 \tag{4.33}$$

$$\frac{U_c}{U_h} = \frac{1}{f_k} - \frac{f_r}{f_k}\rho + \frac{1}{\gamma}(1-\rho) - \frac{nU_G}{U_h}(1-\rho). \tag{4.34}$$

Market equilibrium under the three policies

As in the preceding section, we consider three types of policies to restore social optimality in the decentralized market: a) a DF policy, b) a DR policy, and c) a TB policy. The policy instruments to be considered are once again: a tax levied on consumption (t_c), disposal of waste (t_g), and recycling of waste (t_r).

DF policy

The consumer chooses c, k_h, and ρ to maximize utility,

$$U(c, k_h, G),$$

subject to the budget constraint

$$k - k_h + (p_r - t_r)\,\rho c = (p_c + t_c)\,c + (p_g + t_g)(1 - \rho)\,c.$$

This time labor is numeraire, in contrast to the other model, where it was consumption goods. The price of goods is denoted by p_c. In accordance with the assumption stated above, a part of end-of-life good ρc can be recycled and the rest $(1 - \rho)\,c$ must be disposed of. The price of recyclable parts of a good is denoted by p_r, and the waste management fee for disposal is denoted by p_g.

It is assumed that consumers choose the product attribute represented by ρ.

> The choice is available because competing firms offer different designs (even though the equilibrium with identical households will involve a single outcome for attribute ρ). Individual consumers are price-takers, but in the aggregate they face a "supply" schedule for ρ that is reflected in the price p_c for which firms are willing to sell. If firms devote more resources to "green design," then they will have to charge more for a product with greater recyclability.
>
> (Fullerton and Wu, 1998, pp. 139–140)

Then the FOCs for the consumer's problem are

$$U_h - \mu = 0 \tag{4.35}$$

$$U_c + \mu((p_r - t_r)\,\rho - (p_c + t_c) - (p_g + t_g)(1 - \rho)) = 0 \tag{4.36}$$

$$(p_r - t_r)\,c - \frac{\partial p_c}{\partial \rho}\,c + (p_g + t_g)\,c = 0, \tag{4.37}$$

where the Lagrange multiplier is denoted by μ. From (4.35) and (4.36) we can obtain

$$\frac{U_c}{U_h} = (p_c + t_c) - (p_r - t_r)\,\rho + (p_g + t_g)(1 - \rho). \tag{4.38}$$

Producers choose k_c, r, and ρ to maximize profits, thus:

$$\Pi_c = p_c f(k_c, r, \rho) - p_r r - k_c.$$

Individual firms are price-takers, but in the aggregate they face a 'demand' schedule for ρ that is reflected in the price p_c that consumers are willing to pay. If consumers have to pay for waste disposal, then they will be willing to pay more for a product with greater recyclability ($\partial p_c / \partial \rho \geq 0$)" (Fullerton and Wu, 1998, p. 139).

Then the FOCs become

$$p_c = \frac{1}{f_k} \tag{4.39}$$

$$p_r = \frac{f_r}{f_k} \tag{4.40}$$

$$\frac{\partial p_c}{\partial \rho} c = -\frac{f_\rho}{f_k}. \tag{4.41}$$

Waste management firms maximize $\Pi_g = p_g g - k_g = (p_g \gamma - 1)k_g$. Then we obtain

$$p_g = 1 / \gamma. \tag{4.42}$$

Substituting (4.39), (4.40), and (4.42) into (4.38), we obtain

$$\frac{U_c}{U_h} = \frac{1}{f_k} + t_c - \left(\frac{f_r}{f_k} - t_r\right)\rho + \left(\frac{1}{\gamma} + t_g\right)(1 - \rho). \tag{4.43}$$

Substituting (4.40), (4.41), and (4.42) into (4.37), we obtain

$$-\left(\frac{f_r}{f_k} - t_r\right)c - \frac{f_\rho}{f_k} - \left(\frac{1}{\gamma} + t_g\right)c = 0. \tag{4.44}$$

Under the optimal policy, the competitive market equilibrium conditions (4.43) and (4.44) must coincide with the social optimal conditions (4.33) and (4.34), respectively. If we focus on a DF policy, it can be easily seen that

$$t_g = -nU_G / U_h > 0 \text{ and } t_c = t_r = 0. \tag{4.45}$$

If there is no possibility of illegal disposal, only a disposal tax is necessary and no other policy instruments are needed. This result shows that a DF policy *can* promote the DfE of products, even though the OECD *Guidance Manual* expresses some doubt that a disposal fee can improve the product design (see Section 3.3).

DR policy

If $p_g + t_g > 0$ consumers have a strong incentive to dispose of waste illegally, so we add a policy constraint on the choice of policy instruments, which is represented by $p_g + t_g = 0$. It can easily be seen that the social optimum can be attained if the government sets t_c and t_r such that

$$t_c = \frac{1}{\gamma} - \frac{nU_G}{U_h} > 0 \text{ and} \tag{4.46}$$

$$t_r = -t_c < 0. \tag{4.47}$$

Note that by the constraint stated above $t_g = -\frac{1}{\gamma} < 0$. This policy set is a deposit/refund system. Consumers pay a deposit (t_c) at the purchase stage of goods and they will be paid a refund ($-t_r > 0$) if they sell those end-of-life goods to producers.

TB policy

Under a TB policy, the problems of producers and consumers change. Producers are required to receive end-of-life goods from consumers at no charge to the latter. The producers choose c, r, g, k_c, and ρ to maximize profits,

$$\Pi_c = p_c\, c - k_q - (p_g + t_g)\, q - t_r\, r,$$

subject to $c = f(k_c, r, \rho)$, $r = \rho c$, and $g = (1 - \rho)\, c$. Note that under this policy consumers are indifferent with regard to the recyclability of goods, and consequently producers will not take into account the effect of product recyclability on the price p_c. A simple arrangement of the FOCs yields

$$p_c - \frac{1}{f_k} + \left(\frac{f_r}{f_k} - t_r\right)\rho - (p_g + t_g)\,(1 - \rho) = 0 \tag{4.48}$$

$$\frac{f_\rho}{f_k} + \left(\frac{f_r}{f_k} - t_r\right)c + (p_g + t_g)\, c = 0. \tag{4.49}$$

The consumer chooses c and k_h to maximize utility,

$$U(c, k_h, G),$$

subject to $k - k_h = (p_c + t_c)\, c$.
The FOCs yield

$$\frac{U_c}{U_h} = p_c + t_c. \tag{4.50}$$

Note again that the consumer cannot receive any benefit from product recyclability and consequently will not consider it when purchasing goods. If (4.48) is substituted into (4.50), it can be rewritten as

$$\frac{U_c}{U_h} = t_c + \frac{1}{f_k} - \left(\frac{f_r}{f_k} - t_r\right)\rho + (p_g + t_g)(1 - \rho). \tag{4.51}$$

When we compare the competitive equilibrium conditions (4.49) and (4.51) with the social optimal conditions (4.33) and (4.34), respectively, it must be that

$$t_g = -nU_G / U_h > 0 \text{ and } t_c = t_r = 0. \tag{4.52}$$

This result shows that a TB policy alone cannot attain the social optimum, and that it must be coupled with a disposal fee levied on producers. And yet it has been considered the policy that most purely embodies EPR concepts (see Section 3.4).

Recent discussions regarding DfE

The results derived above presume that consumers will be paid a price for their recyclable items and that the price varies with the degree of product recyclability. However, in the actual economy there is a high transaction cost associated with paying consumers for each individual item to be recycled. Calcott and Walls (2000) show that a DF policy provides incentives for efficient DfE only when there is a fully functioning recycling market. When recycling markets function well, producers have an incentive to lift the recyclability level of their products. The higher the recyclability of their product, the higher the (new) product price is, because the price of end-of-life goods paid to consumers by recyclers increases with the recyclability. This mechanism may not operate in the presence of a high transaction cost within the recycling market.

Calcott and Walls (2005) model two realistic imperfections in recycling markets: transaction costs, and the possibility that consumers incorrectly sort recyclable materials into trash. They assume that consumers have two options for recycling: they may recycle at their curb for free, or they may take items to a recycler and receive payment but incur transaction costs. They examine policies to encourage efficient product design and recycling and show that second-best outcomes can be achieved by combining a deposit/refund system with a modest disposal fee. In the second-best policy, the product tax (i.e., deposit) ensures that the same social cost of disposal is paid for items thrown in the trash. The refund encourages consumers to bring recyclable items for recycling. The disposal fee should be set equal to the social costs of disposal less the difference between recycling costs and virgin material costs.

Eichner and Pethig (2001) define DfE of the product by the material content of the good, or the share of material per unit of good. They consider corrective tax/subsidy schemes if markets fail to reflect material content in their prices. Either the output of the good is taxed and the material input is subsidized, or vice versa, depending on whether an increase or a reduction of material content is to be induced. For example, if the benefit of recycling exceeds the environmental damage, the policy instruments needed for DfE are a tax on the output of the good and a subsidy on the demand for material input.

In a different setting, namely a dynamic durable good model, Eichner and Pethig (2005) derive efficient policies for DfE in the case of free waste disposal. They show that the social optimum can be attained by a deposit/refund system consisting of a subsidy on recyclability of the durable good and a tax on the purchase of the good.

5 The second-best policy for recycling household waste

5.1 Introduction[1]

In Chapter 4 we assumed a first-best economy in which there was no possibility of illegal disposal and showed how the social optimum could be attained by means of a DF policy, or a DR policy, or a TB policy, but because in the real economy illegal disposal is a real possibility, our final conclusion was that EPR policies are preferable to a DF policy.

Now we move on to consider two factors that prevent the first-best outcomes, one being the transaction cost associated with a recycling subsidy (or refund), and the other being illegal disposal by the consumer. We shall see that the social optimum cannot be attained in such a second-best economy.[2] Surprisingly, however, the three policies examined in Chapter 4 come up again as candidates for selection as the second-best policy: a DF policy, a DR policy, and a TB policy.

In the next section we demonstrate that the first-best outcome can be achieved by various combinations of policies, including a disposal fee, a recycling subsidy, and an advance disposal fee. In Section 5.3 the second-best policy is characterized by the presence of two factors that stand in the way of the first-best optimum: illegal disposal and the transaction costs associated with a recycling subsidy.

5.2 The first-best policy

Let us consider two kinds of goods. First, firm 1 produces good 1 by labor input l_1 and its production function is represented by $f(l_1)$. We assume that good 1 is totally consumed and generates no waste after consumption.

Second, firm 2 produces good 2 from labor input l_2 and end-of-life good 2. The amount of end-of-life good 2 to be recycled is denoted by r. Therefore, we assume that good 2 is recyclable but firm 2 intermediately has to treat collected good 2 for recycling before using it as an input in the production process, in which case the cost associated with the intermediate treatment for recycled input is represented by cr. The production function of good 2 is represented by $g(l_2 + \beta r)$. Both c and β are assumed to be constant. Alternatively, good 2 can be regarded as made from a primary resource and a secondary resource, if it is assumed that one unit of the primary resource can be extracted by one unit of labor. End-of-life good 2 would be waste if it is not recycled. The constant marginal waste disposal cost is denoted by γ.

We assume that a representative consumer is endowed with the total amount of labor L. The utility function of the consumer is represented by $U(q_1, q_2)$ where q_1 and q_2 denote the consumption of good 1 and good 2, respectively. The consumer decides not only the amount of consumption of goods but also the amount of good 2 prepared for recycling after consumption. If the consumer wants to recycle consumed good 2, he or she is required to prepare the goods for recycling by, for example, separating them from other things and cleaning them. The preparation cost is represented by $\theta(r)$, with $\theta'(r) > 0$ and $\theta''(r) > 0$.

In this section we first of all present a competitive equilibrium model including an advance disposal fee, a disposal fee, and a recycling subsidy. We then analyze the first-best policy generating the social optimal outcome.

We choose labor as numeraire and therefore the wage rate is 1. The profits of firm 1, π_1, and that of firm 2, Π_2, are represented by $\Pi_1 = P_1 f(l_1) - l_1$ and $\Pi_2 = (P_2 - t)g(l_2 + \beta r) - cr - sr - l_2$ respectively. The prices of good 1 and good 2 are denoted by P_1 and P_2 respectively. The price of end-of-life good 2 prepared for recycling is denoted by s, which is paid by firm 2 to the consumer. Note that the price s can be negative. The advance disposal fee is denoted by t. Labor input l_1 and l_2 can be written as functions of q_1 and q_2 using the corresponding production functions as follows:

$$l_1 = f^{-1}(q_1) \text{ and}$$

$$l_2 = g^{-1}(q_2) - \beta r,$$

where f^{-1} and g^{-1} represent the inverse functions of f and g respectively. Using these relations the profits can be written as $\Pi_1 = P_1 q_1 - f^{-1}(q_1)$ and $\Pi_2 = (P_2 - t)q_2 - cr - sr - g^{-1}(q_2) + \beta r$. The owner of firm 1 is supposed to maximize the profit by choosing q_1 instead of l_1. Similarly, the owner of firm 2 is supposed to maximize the profit by choosing q_2 and r instead of l_2 and r.

The first-order conditions for firms are given by:

$$P_1 - \frac{1}{f'} = 0 \tag{5.1}$$

$$(P_2 - t) - \frac{1}{g'} = 0 \tag{5.2}$$

$$\beta - c - s = 0. \tag{5.3}$$

The consumer is confronted with a disposal fee denoted by τ, if he or she disposes of good 2 after consumption. The consumer, however, can acquire a recycling subsidy denoted by ϕ, if he or she prepares consumed good 2 for recycling.

The consumer chooses q_1, q_2, and r so as to maximize $U(q_1, q_2)$ subject to his or her budget constraint. The budget constraint is described as:

$$\Pi - P_1 q_1 - P_2 q_2 - \theta(r) + (s + \phi)r - \tau(q_2 - r) = 0, \tag{5.4}$$

where Π is the sum of the profits of firms and the revenue of the municipality, $(s + \phi)r$ is the sum of rewards for recycling activity, and $\tau(q_2 - r)$ is the disposal fee. Lagrangian Ω can be represented by:

$$\Omega = U(q_1, q_2) + \lambda\{\Pi - P_1 q_1 - P_2 q_2 - \theta(r) + (s + \phi)r - \tau(q_2 - r)\}, \tag{5.5}$$

where λ denotes the Lagrangian multiplier. The first-order conditions for the representative consumer are represented by:

$$U_1 - \lambda P_1 = 0 \tag{5.6}$$

$$U_2 - \lambda(P_2 + \tau) = 0 \tag{5.7}$$

$$-\theta' + (s + \phi) + \tau = 0. \tag{5.8}$$

Using equations (5.1) to (5.3), first-order conditions for the consumer (5.6) to (5.8) can be rewritten as follows:

$$U_1 - \lambda \frac{1}{f'} = 0 \tag{5.9}$$

$$U_2 - \lambda(\frac{1}{g'} + t + \tau) = 0 \tag{5.10}$$

$$\beta - c - \theta' + \phi + \tau = 0. \tag{5.11}$$

Equation (5.9) is obtained from (5.1) and (5.6) by eliminating P_1. Equation (5.10) is derived from (5.2) and (5.7) by eliminating P_2. We can obtain (5.11) from (5.3) and (5.8) by eliminating s.

Equations (5.4), (5.9), (5.10), and (5.11) give the competitive equilibrium values of q_1, q_2, r, and λ, all of which depend on t, τ, and ϕ. Substituting the equilibrium values into the utility function, we can obtain the maximized utility function, $U(q_1^* (t, \tau, \phi), q_2^* (t, \tau, \phi))$, where the asterisk denotes the competitive equilibrium values. The social optimal level for t, τ, and ϕ should be set to maximize the maximized utility function. The envelope theorem, however, ensures the following relation,

$$\frac{\partial U(q_1^* (t, \tau, \phi), q_2^* (t, \tau, \phi))}{\partial \alpha} = \frac{\partial \Omega(q_1^* (t, \tau, \phi), q_2^* (t, \tau, \phi), r^* (t, \tau, \phi), \lambda^* (t, \tau, \phi))}{\partial \alpha} \Big|_{\alpha = t,\, \tau,\, \phi}.$$

Substituting the equilibrium values into (5.5), we can obtain $\Omega(q_1^* (t, \tau, \phi), q_2^* (t, \tau, \phi), r^* (t, \tau, \phi), \lambda^* (t, \tau, \phi))$ as follows:

$$\Omega(q_1^*, q_2^*, r^*, \lambda^*) = U(q_1^*, q_2^*) + \lambda^* \{L - f^{-1}(q_1^*) - g^{-1}(q_2^*) + \beta r^*$$

$$- cr^* - \theta(r^*) - \gamma(q_2^* - r^*)\}.$$

Therefore, the first-order conditions for the optimal policy intervention (choice of t, τ, and ϕ) are given by:

$$\frac{\partial\Omega^*}{\partial q_1^*}\frac{\partial q_1^*}{\partial\alpha} + \frac{\partial\Omega^*}{\partial q_2^*}\frac{\partial q_2^*}{\partial\alpha} + \frac{\partial\Omega^*}{\partial r^*}\frac{\partial r^*}{\partial\alpha} + \frac{\partial\Omega^*}{\partial\lambda^*}\frac{\partial\lambda^*}{\partial\alpha} = 0,\ \alpha = t,\ \tau, \text{ or } \phi \tag{5.12}$$

We can easily calculate $\frac{\partial\Omega^*}{\partial q_1^*}$, $\frac{\partial\Omega^*}{\partial q_2^*}$, $\frac{\partial\Omega^*}{\partial r^*}$, and $\frac{\partial\Omega^*}{\partial\lambda^*}$ as follows:

$$\frac{\partial\Omega^*}{\partial q_1^*} = U_1 - \lambda^* \frac{1}{f'}$$

$$\frac{\partial\Omega^*}{\partial q_2^*} = U_2 - \lambda^* \left(\frac{1}{g'} + \gamma\right)$$

$$\frac{\partial\Omega^*}{\partial r^*} = \lambda^* (\beta - c - \theta' + \gamma)$$

$$\frac{\partial\Omega^*}{\partial\lambda^*} = L - f^{-1} - g^{-1} + \beta r^* - cr^* - \theta(r^*) - \gamma(q_2^* - r^*).$$

We can see that $\frac{\partial\Omega^*}{\partial q_1^*}$ and $\frac{\partial\Omega^*}{\partial\lambda^*}$ are always zero by (5.9) and (5.4) respectively. It follows from (5.10) and (5.11) that $\frac{\partial\Omega^*}{\partial q_2^*}$ and $\frac{\partial\Omega^*}{\partial r^*}$ can be zero by setting both $t + \tau$ and $\tau + \phi$ to γ. Therefore, in the first-best economy, the optimal policy intervention can be characterized by:

$$t + \tau = \gamma \tag{5.13}$$

$$\phi + \tau = \gamma. \tag{5.14}$$

Equation (5.13) shows that the first-best optimum can be achieved in the decentralized market through various combinations of t and τ such that $t + \tau = \gamma$. Equation (5.14) shows that either or both a disposal fee and a recycling subsidy are necessary in order to give the consumer the incentive to recycle consumed good 2.

From equations (5.13) and (5.14) we can see that the first-best outcomes can be obtained by the three policies examined in the previous chapter, i.e., a DF policy, a DR policy, or a TB policy. First, a DF policy is one of the first-best policies, when τ is set to γ. Second, a DR policy is available for achieving the first-best optimum, where both deposit (= t) and refund (= ϕ) are set to γ without a disposal fee.

Third, the first-best outcome can also be achieved by a TB policy. In a TB policy the consumer is required to prepare for recycling (or to pay a preparation cost $\theta(r)$) without any reward ($s = \phi = 0$). The collected-for-recycling good 2 is received by firm 2. The first-best outcome can also be obtained by such a producer take-back requirement system as characterized by $t = c - \beta$ and $\tau = \beta - c + \gamma$.[3]

5.3 The second-best policy

5.3.1 Two factors that prevent the first-best optimum

In the real economy, however, various factors prevent the first-best outcome. One of them is the transaction cost associated with a recycling subsidy. The consumer must go through various kinds of procedures in order to receive a recycling subsidy and as a result an associated administrative cost will be generated. The constant marginal transaction cost associated with a recycling subsidy is supposed to be π. Therefore, the consumer is required to pay $\theta(r) + \pi r$ instead of $\theta(r)$. For simplicity, we assume $\beta - c + \gamma > \pi$, which ensures an interior solution.

A second factor preventing the first-best outcome is illegal disposal. The consumer is tempted to illegally dispose of consumed good 2, if he or she is confronted with a disposal fee. The consumer may throw rubbish by the side of the road or take it to the nearest public waste receptacles, such as rubbish bins in a park. The quantity of illegal disposal is denoted by d. Following along the lines adopted by Fullerton and Kinnaman (1995), we assume that the consumer is to pay the cost associated with illegal disposal, which is represented by $v(d)$, with $v' > 0$ and $v'' > 0$. In addition, it will cost society, as a whole, to collect illegally disposed of good 2 and return it to the proper disposal process. Denoting the constant marginal treatment cost associated with illegal disposal by γ', we assume $\gamma' > \gamma$.

The budget constraint of the consumer should be replaced by:

$$\Pi - P_1 q_1 - P_2 q_2 - \theta(r) + (s + \phi) r - \pi r - \tau (q_2 - r - d) - v(d) = 0. \tag{5.15}$$

Note that the amount of rubbish legally disposed of decreases from $q_2 - r$ to $q_2 - r - d$ because of illegal disposal. Similarly, the Lagrangian for consumer's problem (5.5) should be replaced by:

$$\Omega = U(q_1, q_2) + \lambda \{\Pi - P_1 q_1 - P_2 q_2 - \theta(r) + (s + \phi) r - \pi r - \tau(q_2 - r - d) - v(d)\}. \tag{5.16}$$

The first-order condition associated with recycling should be replaced by the following equation instead of (5.11):

$$\beta - c - \theta' - \pi + \phi + \tau = 0. \tag{5.17a}$$

A new first-order condition associated with illegal disposal should also be added:

$$-v' + \tau = 0. \tag{5.18}$$

Note that other first-order conditions (5.9) and (5.10) remain unchanged even in the second-best economy. Equations (5.17a) and (5.18) mean that r is a function of $\tau + \phi$ and d is a function solely of τ.

We can dispense with a recycling subsidy in a TB policy, which is also a candidate for the second-best policy. However, in order to give the consumer the incentive to recycle, we need a disposal fee and therefore we have to consider the problem of illegal disposal.

The first-order condition associated with recycling should be replaced by the following equation, in the TB policy, instead of (5.17a):

$$-\theta' + \tau = 0, \tag{5.17b}$$

where note that $s = \phi = \tau = 0$ in (5.15) and (5.16). The first-order condition associated with illegal disposal (5.18) is still valid.

5.3.2 *Three candidates for the second-best policy*

First we consider a mixed policy consisting of a disposal fee, an advance disposal fee, and a recycling subsidy. Substituting the competitive equilibrium values in the second-best economy into Lagrangian (5.16) in a similar way as in the previous section, the maximized Lagrangian can be obtained:

$$\begin{aligned}
&\Omega(q_1^*, q_2^*, r^*, \lambda^*)\\
&= U(q_1^*, q_2^*) + \lambda^* \{L - f^{-1}(q_1^*) - g^{-1}(q_2^*) + \beta r^* - cr^* - \theta(r^*)\\
&- \gamma(q_2^* - r^*) - v(d^*) - (\gamma' - \gamma)d^*\},
\end{aligned}$$

where q_1^*, q_2^*, r^*, d^*, and λ^* depend on, t, τ, and ϕ.

We choose t, ϕ, and τ so as to maximize Ω^*. The first-order conditions are given by:

$$\frac{\partial \Omega^*}{\partial q_1^*}\frac{\partial q_1^*}{\partial t} + \frac{\partial \Omega^*}{\partial q_2^*}\frac{\partial q_2^*}{\partial t} + \frac{\partial \Omega^*}{\partial \lambda^*}\frac{\partial \lambda^*}{\partial t} + \frac{\partial \Omega^*}{\partial r^*}\frac{\partial r^*}{\partial t} + \frac{\partial \Omega^*}{\partial d^*}\frac{\partial d^*}{\partial t} = 0 \tag{5.19}$$

$$\frac{\partial \Omega^*}{\partial q_1^*}\frac{\partial q_1^*}{\partial \phi} + \frac{\partial \Omega^*}{\partial q_2^*}\frac{\partial q_2^*}{\partial \phi} + \frac{\partial \Omega^*}{\partial \lambda^*}\frac{\partial \lambda^*}{\partial \phi} + \frac{\partial \Omega^*}{\partial r^*}\frac{\partial r^*}{\partial \phi} + \frac{\partial \Omega^*}{\partial d^*}\frac{\partial d^*}{\partial \phi} = 0 \tag{5.20}$$

$$\frac{\partial \Omega^*}{\partial q_1^*}\frac{\partial q_1^*}{\partial \tau} + \frac{\partial \Omega^*}{\partial q_2^*}\frac{\partial q_2^*}{\partial \tau} + \frac{\partial \Omega^*}{\partial \lambda^*}\frac{\partial \lambda^*}{\partial \tau} + \frac{\partial \Omega^*}{\partial r^*}\frac{\partial r^*}{\partial \tau} + \frac{\partial \Omega^*}{\partial d^*}\frac{\partial d^*}{\partial \tau} = 0. \tag{5.21}$$

Similarly, we can easily calculate $\frac{\partial \Omega^*}{\partial q_1^*}$, $\frac{\partial \Omega^*}{\partial q_2^*}$, $\frac{\partial \Omega^*}{\partial r^*}$, $\frac{\partial \Omega^*}{\partial d^*}$, and $\frac{\partial \Omega^*}{\partial \lambda^*}$, as follows:

$$\frac{\partial \Omega^*}{\partial q_1^*} = U_1 - \lambda^* \frac{1}{f'}$$

$$\frac{\partial \Omega^*}{\partial q_2^*} = U_2 - \lambda^* \left(\frac{1}{g'} + \gamma\right)$$

$$\frac{\partial \Omega^*}{\partial r^*} = \lambda^* (\beta - c - \theta' - \pi + \gamma)$$

$$\frac{\partial \Omega^*}{\partial d^*} = -\lambda^* (v'(d^*) + \gamma' - \gamma)$$

$$\frac{\partial \Omega^*}{\partial \lambda^*} = L - f^{-1} - g^{-1} + \beta r^* - cr^* - \theta(r^*) - \pi r^* - \gamma(q_2^* - r^*) - v(d^*)$$

$$-(\gamma' - \gamma)d^*.$$

It can be seen that $\frac{\partial \Omega^*}{\partial q_1^*}$ and $\frac{\partial \Omega^*}{\partial \lambda^*}$ are always zero by (5.9) and (5.15) respectively. Similarly, $\frac{\partial r^*}{\partial t}$ and $\frac{\partial d^*}{\partial t}$ are also zero, since r is a function of $\tau + \phi$ and d is a function solely of τ, which follow from (5.17a) and (5.18). Therefore, it follows from (5.19) that $\frac{\partial \Omega^*}{\partial q_2^*}$ should be zero. Note that $\frac{\partial \Omega^*}{\partial q_2^*}$ can be zero by setting $t + \tau$ to γ. Therefore, the first-order conditions through (5.19) to (5.21) can be rewritten as:

$$t + \tau = \gamma \tag{5.22}$$

$$\frac{\partial \Omega^*}{\partial r^*}\frac{\partial r^*}{\partial \phi} + \frac{\partial \Omega^*}{\partial d^*}\frac{\partial d^*}{\partial \phi} = 0 \tag{5.23}$$

$$\frac{\partial \Omega^*}{\partial r^*}\frac{\partial r^*}{\partial \tau} + \frac{\partial \Omega^*}{\partial d^*}\frac{\partial d^*}{\partial \tau} = 0. \tag{5.24}$$

Equations (5.17a) and (5.18) imply $\frac{\partial r^*}{\partial \phi} = \frac{\partial r^*}{\partial \tau} > 0$, $\frac{\partial d^*}{\partial \phi} = 0$, and $\frac{\partial d^*}{\partial \tau} > 0$. Note that $\frac{\partial \Omega^*}{\partial d^*}$ is negative. Therefore, the left-hand side of (5.24) is always smaller than that of (5.23). Equivalently, $\frac{\partial \Omega^*}{\partial \phi}$ is larger than $\frac{\partial \Omega^*}{\partial \tau}$ and both of them cannot be zero simultaneously, which means that the policy-maker should not apply a recycling subsidy and a disposal fee simultaneously.

If a recycling subsidy is adopted, the transaction cost πr should be paid regardless of how much recycling subsidy is given to the consumer. Equation (5.17a), however, implies that ϕ is a complete substitute for τ. Therefore, once a recycling subsidy has been adopted, it becomes rational to abandon a disposal fee. This is an intuitive reason for why a second-best policy consists of either a recycling subsidy or a disposal fee.

First, suppose that a recycling subsidy will be applied instead of a disposal fee. The first-order condition (5.23) can be simplified as:

$$(\beta - c - \theta' - \pi + \gamma)\frac{\partial r}{\partial \phi} = 0. \tag{5.25}$$

It follows from (5.17a) that (5.25) can be satisfied by setting ϕ to γ. Therefore, a recycling subsidy policy is characterized by $t = \phi = \gamma$, which is a DR policy.

Second, we alternatively assume choosing a disposal fee. The first-order condition for τ (5.24) can be rewritten as:

$$(\beta - c - \theta' + \gamma)\frac{\partial r^*}{\partial \tau} - (v' + \gamma' - \gamma)\frac{\partial d^*}{\partial \tau} = 0. \tag{5.26}$$

Note that π is eliminated by giving up a recycling subsidy. According to this first-order condition, the disposal fee should be set such that the marginal social benefit obtained from increasing one unit of τ is equal to the associated marginal social cost. The first term of the left-hand side of equation (5.26) represents the marginal net benefit from recycling activity promoted by increasing one unit of τ, and the second term of the equation represents the marginal net cost from the illegal disposal activity of the consumer tempted by increasing one unit of τ. We can obtain the derivative of r^* and d^* with respect to τ by (5.17a) and (5.18) respectively as follows:

$$\frac{\partial r^*}{\partial \tau} = \frac{1}{\theta''}$$

$$\frac{\partial d^*}{\partial \tau} = \frac{1}{v''}.$$

Substituting them into (5.26), we can obtain the following equation:

$$\frac{\beta - c - \theta' + \gamma}{\theta''} - \frac{v' + \gamma' - \gamma}{v''} = 0. \tag{5.27}$$

Equations (5.17a), (5.18), and (5.27) define implicitly the optimal value of τ. Therefore, the DF policy, which is also a candidate for the second-best policy, is characterized by (5.17a), (5.18), (5.22), and (5.27).

Next, we can consider another candidate for the second-best policy, a TB policy. Note that equation (5.27) is still valid in the TB policy, since (5.17b) and (5.18) ensure $\frac{\partial r^*}{\partial \tau} = \frac{1}{\theta''}$ and $\frac{\partial d^*}{\partial \tau} = \frac{1}{v''}$. Therefore, the TB policy is characterized by (5.17b), (5.18), (5.22), and (5.27).

What is the second-best policy? Is it the DF policy with an advance disposal fee, the DR policy, or the TB policy? Equation (5.25) is also the first-order condition for maximizing $(\beta - c + \gamma - \pi)r - \theta(r)$. Similarly, (5.27) is also the first-order condition for maximizing $(\beta - c + \gamma)r - \theta(r) - v(d) - (\gamma' - \gamma)d$. It is, therefore, the second-best policy that gives the maximum of $(\beta - c - \pi + \gamma)r - \theta(r)$ and $(\beta - c + \gamma)r - \theta(r) - v(d) - (\gamma' - \gamma)d$, where the former corresponds to the DR policy and the latter corresponds to the DF policy with an advance disposal fee and the TB policy.

5.3.3 *The characterization of the second-best policy in a parametric space*

In this section, we derive the second-best policy. Note that the price of the collected-for-recycling good 2, $s = \beta - c$, is also regarded as a parameter. Therefore, we can characterize the second-best policy in the $(\pi - s)$-plane.

We define *DF* and *TB* respectively as the maximum value of $(s + \gamma)r - \theta(r) - v(d) - (\gamma' - \gamma)d$ under the corresponding policies, and we also define *DR* as the maximum value of $(s - \pi + \gamma)r - \theta(r)$ under the corresponding policy:

$$DF = (s + \gamma)r_{df}(\tau_{df}^*, s) - \theta(r_{df}(\tau_{df}^*, s)) - v(d_{df}(\tau_{df}^*)) - (\gamma' - \gamma)d_{df}(\tau_{df}^*)$$

$$TB = (s + \gamma)r_{tb}(\tau_{tb}^*, s) - \theta(r_{tb}(\tau_{tb}^*, s)) - v(d_{tb}(\tau_{tb}^*)) - (\gamma' - \gamma)d_{tb}(\tau_{tb}^*)$$

$$DR = (s - \pi + \gamma)r_{dr}(\phi_{dr}^*, s, \pi) - \theta(r_{dr}(\phi_{dr}^*, s, \pi)).$$

Note that $r_{df}(\tau_{df}^*, s)$ and $d_{df}(\tau_{df}^*)$ are derived from (5.17a) and (5.18) respectively, that $r_{tb}(\tau_{tb}^*, s)$ and $d_{tb}(\tau_{tb}^*)$ are from (5.17b) and (5.18) respectively, and that $r_{dr}(\phi_{dr}^*, s, \pi)$ is from (5.17a). The first-order condition for the optimal value of τ_{up} and τ_{tb} is (5.27). Note that the optimal value of ϕ_{dr} is γ. $\phi_{dr}^* = \gamma$.

Using the envelope theorem we can obtain the derivatives of those maximum value functions with respect to s and π.

With respect to DF

We can obtain by the envelope theorem the following result:

$$\frac{\partial DF}{\partial s} = r_{df}^* + (s + \gamma - \theta'(r_{df}^*))\frac{\partial r_{df}^*}{\partial s} = r_{df}^* + (\gamma - \tau_{df}^*)\frac{\partial r_{df}^*}{\partial s}. \tag{5.28}$$

The partial derivative $\frac{\partial r_{df}^*}{\partial s}$ can be calculated from (5.17a):

$$\frac{\partial r_{df}^*}{\partial s} = \frac{1}{\theta''}. \tag{5.29}$$

It should be noted from (5.17a) that $\tau_{df}^* < \gamma$. Therefore, the sign of $\frac{\partial DF}{\partial s}$ is positive. Similarly, we can obtain the derivative of *DF* with respect to π:

$$\frac{\partial DF}{\partial \pi} = 0. \tag{5.30}$$

With respect to TB

The envelope theorem ensures the following result:

$$\frac{\partial TB}{\partial s} = r_{tb}^* + (s + \gamma - \theta'(r_{tb}^*))\frac{\partial r_{tb}^*}{\partial s} = r_{tb}^*. \tag{5.31}$$

Note that the partial derivative $\frac{\partial r_{tb}^*}{\partial s}$ is zero by (5.17b). Similarly, we can obtain the derivative of *TB* with respect to π:

$$\frac{\partial TB}{\partial \pi} = 0. \tag{5.32}$$

With respect to DR

By the envelope theorem, the following results can be obtained:

$$\frac{\partial DR}{\partial s} = r_{dr}^* + (s - \pi + \gamma - \theta'(r_{dr}^*))\frac{\partial r_{dr}^*}{\partial s} = r_{dr}^* \tag{5.33}$$

$$\frac{\partial DR}{\partial \pi} = -r_{dr}^* + (s - \pi + \gamma - \theta'(r_{dr}^*))\frac{\partial r_{dr}^*}{\partial \pi} = -r_{dr}^*. \tag{5.34}$$

Note that $s - \pi + \gamma - \theta'(r_{dr}^*)$ is always zero as long as ϕ_{dr}^* is fixed to γ.

Before characterizing the second-best policy, it is convenient to derive the total derivatives of r and τ with respect to s. Using (5.17a), (5.18), and (5.27), we can calculate the total derivatives dr_{df}^*/ds and $d\tau_{df}^*/ds$ as follows:

$$\frac{dr_{df}^*}{ds} = \frac{v''(\theta'' + v'') - (s - \theta' + \gamma)v'''}{(\theta''v'' + (v' + \gamma' - \gamma)\theta''')v'' + (\theta''v'' - (s - \theta' + \gamma)v''')\theta''}$$

$$\frac{d\tau_{df}^*}{ds} = \frac{\theta''v''v'' - (\theta''v'' + (v' + \gamma' - \gamma)\theta''')v''}{(\theta''v'' + (v' + \gamma' - \gamma)\theta''')v'' + (\theta''v'' - (s - \theta' + \gamma)v''')\theta''}$$

Assuming that θ''' and v''' are sufficiently small to be neglected, $\frac{dr_{df}^*}{ds}$ and $\frac{d\tau_{df}^*}{ds}$ can be rewritten as:

$$\frac{dr_{df}^*}{ds} = \frac{1}{\theta''} \tag{5.35}$$

$$\frac{d\tau_{df}^*}{ds} = 0. \tag{5.36}$$

Trying comparative statics among (5.17b), (5.18), and (5.27), we can calculate $\frac{dr_{tb}^*}{ds}$ and $\frac{d\tau_{tb}^*}{ds}$ as follows:

$$\frac{dr_{tb}^*}{ds} = \frac{v''v''}{(\theta''v'' + (v' + \gamma' - \gamma)\theta''')v'' + (\theta''v'' - (s - \theta' + \gamma)v''')\theta''}$$

$$= \frac{v''}{\theta''(\theta'' + v'')} \tag{5.37}$$

$$\frac{d\tau_{tb}^*}{ds} = \frac{\theta''v''v''}{(\theta''v'' + (v' + \gamma' - \gamma)\theta''')v'' + (\theta''v'' - (s - \theta' + \gamma)v''')\theta''}$$

$$= \frac{v''}{\theta'' + v''}, \tag{5.38}$$

assuming that θ''' and v''' are sufficiently small to be neglected.

Now we are ready to characterize the second-best policy. First, we compare the DF policy with the TB policy. In order to compare these policies, we have to see the sign of Δ_1, which is defined as:

$$\Delta_1 = DF - TB.$$

We can show that Δ_1 is also positive when s is positive and that Δ_1 is also negative when s is negative. To see this, note that the derivative of Δ_1 with respect to s can be written as:

$$\frac{\partial \Delta_1}{\partial s} = r_{df}^* + (\gamma - \tau_{df}^*)\frac{\partial r_{df}^*}{\partial s} - r_{tb}^*, \tag{5.39}$$

using (5.28) and (5.31).

Now we assume $s = 0$. Then we can see that equation (5.17a) coincides with equation (5.17b), and it follows that $\tau_{df}^* = \tau_{tb}^*$, $r_{df}^* = r_{tb}^*$, and $d_{df}^* = d_{tb}^*$. Therefore, Δ_1 is also zero if the price of the collected-for-recycling good 2, s, is zero. Note that, since $\frac{\partial r_{df}^*}{\partial s} > 0$ (by (5.29)), the sign of $\frac{\partial \Delta_1}{\partial s}$ is positive when s is zero. In addition, we can obtain the following inequality:

$$\frac{\partial^2 \Delta_1}{\partial s^2} = \frac{d}{ds}(r_{df}^* - r_{tb}^*) + (\gamma - \tau_{df}^*)\frac{d\left(\frac{\partial r_{df}^*}{\partial s}\right)}{ds} - \frac{d\tau_{df}^*}{ds}\frac{\partial r_{df}^*}{\partial s}$$
$$= \left(\frac{1}{\theta''}\right) - \left(\frac{v''}{\theta''(\theta'' + v'')}\right) > 0. \tag{5.40}$$

The second equality of (5.40) follows form (5.29), (5.35), (5.36), and (5.37). Therefore, it can be concluded that $DF > TB$ when $s > 0$. The inequality (5.40) indicates that Δ_1 can be positive again when s becomes negative and sufficiently large in the absolute value. This possibility, however, can be excluded. To see this, suppose that $s = -\gamma$. Then we obtain the corner solution represented by, $r_{df}^* = r_{tb}^* = 0$, $\tau_{df}^* = \tau_{tb}^* = 0$, and $d_{df}^* = d_{tb}^* = 0$, which ensure that Δ_1 is also zero. Note that it is assumed that $s + \gamma > \pi$. Therefore, Δ_1 remains negative in the range of s represented by $(\pi - \gamma, 0)$.

It follows that the DF policy is superior to the TB policy if the price of the collected-for-recycling good 2, s, is positive and vice versa. In the DF policy, τ_{df}^* should be smaller than γ, since it tempts the consumer to undertake illegal disposal activity. Therefore, the amount of the recycled good must be smaller in the second-best economy than in the first-best economy. Note that $r_{df}^* = r_{tb}^*$ if the price of the collected-for-recycling good is zero. It follows from (5.35) and (5.37) that $r_{df}^* > r_{tb}^*$ if the price of the collected-for-recycling good is positive, $s > 0$, and vice versa. This is an intuitive reason why the DF policy is superior to the TB policy if $s > 0$.

Second, we compare the DR policy with the DF policy. We have to see the sign of Δ_2, which is defined as:

$$\Delta_2 = DR - DF.$$

We define the critical value of the marginal transaction cost, π, π^*, as follows: if the actual value of π is greater than the critical value π^*, the DF policy is superior to the DR policy. Using (5.28), (5.30), (5.33), and (5.34), the derivative of π^*, with respect to s can be written as:

$$\frac{\partial \pi^*}{\partial s} = \frac{r_{dr}^* - r_{df}^* - (\gamma - \tau_{df}^*)\dfrac{\partial r_{df}^*}{\partial s}}{r_{dr}^*}. \tag{5.41}$$

Unless the value of r_{dr}^* is sufficiently larger than r_{df}^*, the sign of $\frac{\partial \pi^*}{\partial s}$ is negative. Generally, however, the sign of the right-hand side of (5.41) cannot be determined.

Third, we compare the DR policy with the TB policy. We have to see the sign of Δ_3, which is defined as:

$$\Delta_3 = DR - TB.$$

Similarly, we define the critical value of the marginal transaction cost, π, π^{**}, as follows: if the actual value of π is greater than the critical value π^{**}, the TB policy is superior to the DR policy. Using (5.31), (5.32), (5.33), and (5.34), we can derive the derivative of π^{**} with respect to s:

$$\frac{\partial \pi^{**}}{\partial s} = \frac{r_{dr}^* - r_{tb}^*}{r_{dr}^*}. \tag{5.42}$$

If the value of r_{dr}^* is larger than r_{tb}^*, the sign of $\frac{\partial \pi^{**}}{\partial s}$ is positive. Generally, we cannot decide the sign of the right-hand side of (5.42).

Both π^* and π^{**} can be regarded as functions of s. Note that π^* and π^{**} intersect each other on the line of $s=0$ in the $(\pi - s)$-plane, since *DF* and *TB* have a common value when s is zero. The sign of $\frac{\partial \pi^*}{\partial s}$ and $\frac{\partial \pi^{**}}{\partial s}$ depend on how relatively large the value of r^* is among the three policies as stated above. However, it can be shown that $\frac{\partial \pi^{**}}{\partial s} > \frac{\partial \pi^*}{\partial s}$, although we cannot be sure whether π^* and π^{**} are increasing or decreasing functions of s. Alternatively, suppose that $\frac{\partial \pi^{**}}{\partial s} < \frac{\partial \pi^*}{\partial s}$. Figure 5.1 illustrates the π^* and π^{**} that satisfy this supposition. In the shaded area *A* in the same figure, the following inequalities are satisfied: *TB*>*DR*, *DR*>*DF*, and *DF* > *TB*. This is a contradiction. Similarly, in the shaded area *B* in the figure, the transitivity condition is also violated, since inequalities *DF*>*DR*, *DR*>*TB*, and *TB*>*DF* are satisfied in this area. This is also a contradiction. Therefore, the inequality $\frac{\partial \pi^{**}}{\partial s} > \frac{\partial \pi^*}{\partial s}$ should be satisfied.

An example compatible with the inequality, $\frac{\partial \pi^{**}}{\partial s} > \frac{\partial \pi^*}{\partial s}$, is illustrated by Figure 5.2. The whole area in Figure 5.2 is split into three areas corresponding to the particular second-best policy. Note that the figure includes the area represented by $s + \gamma > \pi$, which is an assumption.

It is worth stating the results obtained in this section in the form of propositions.

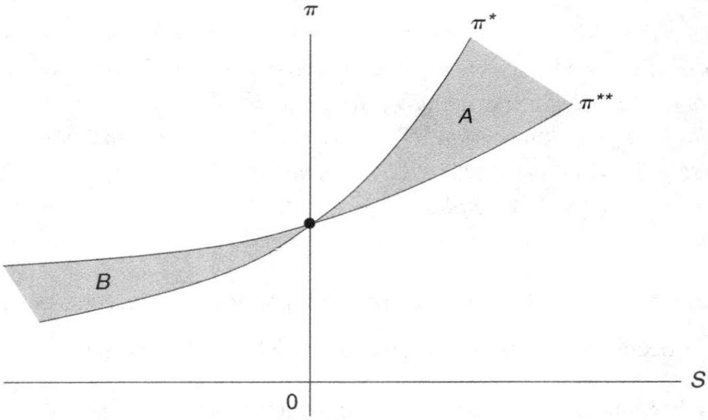

Figure 5.1 The critical values of the marginal transaction cost; π^{*} and π^{**}.

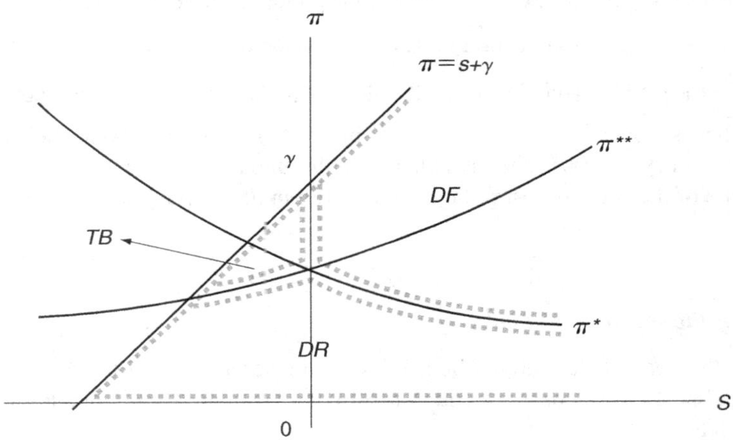

Figure 5.2 The second-best policy and the corresponding area in the $(\pi - s)$-plane.

Proposition 5.1. *If the price of the collected-for-recycling good is positive, a DF policy with an advance disposal fee is superior to a TB policy. Conversely, if the price of the collected-for-recycling good is negative, a TB policy is superior to a DF policy with an advance disposal fee.*

Proposition 5.2. *(i) Assume that the price of the collected-for-recycling good is positive. In addition, if the actual marginal transaction cost* π *is lower than* π^{*}*, a DR policy is the second-best policy. Conversely, if the* *actual marginal transaction cost π is higher than π^*, a DF policy with an advance disposal fee is the second-best policy. (ii) Assume that the price of the collected-for-recycling good is negative. In addition, if the actual marginal transaction cost π is lower than π^{**}, a DR policy is the second-best policy. Conversely, if the actual marginal transaction cost π is higher than π^{**}, a TB policy is the second-best policy.*

There is another criterion for deciding exactly which policy is the second-best. If the sign of $\frac{\partial \pi^*}{\partial s}$ becomes negative, the area in which the DF policy is the second-best policy becomes larger. According to equation (5.41), the sign of $\frac{\partial \pi^*}{\partial s}$ is negative, unless the value of r_{dr}^* is sufficiently larger than r_{df}^*. Similarly, if the sign of $\frac{\partial \pi^{**}}{\partial s}$ becomes positive, the area in which the TB policy is the second-best policy becomes larger. According to equation (5.42), if the value of r_{dr}^* is larger than r_{tb}^*, the sign of $\frac{\partial \pi^{**}}{\partial s}$ is positive and vice versa. It follows that the amount recycled in the associated policy is crucially important. The possibility of the associated policy being the second-best policy increases directly in line with the amount recycled.

An example of the second-best policy is presented in the Appendix.

Appendix

Illustration of the second-best policy

Assume that the second derivatives both of recycling preparation cost θ'' and illegal disposal cost v'' are constant, and that $\theta(r)$ and $v(r)$ take simple quadratic functions such as:

$$\theta(r) = \frac{\vartheta}{2} r^2$$

$$v(d) = \frac{V}{2} d^2$$

in order to illustrate the second-best policy, where ϑ and V are constant.

In the DR policy, we can obtain the optimal quantity of recycling by solving (5.25) for r:

$$r_{dr}^* = \frac{s + \gamma - \pi}{\vartheta}. \tag{A.5.1}$$

In the case of the DF policy with an advance disposal fee, we can obtain the optimal unit pricing level τ_{df}^* by (5.17a), (5.18), and (5.27):

$$\tau_{df}^* = \gamma - \frac{\vartheta}{V + \vartheta} \gamma'. \tag{A.5.2}$$

Solving equations (5.17a), (5.18), and (5.27) for r and d, respectively, we obtain:

$$r_{df}^* = \frac{s + \tau_{df}^*}{\vartheta} = \frac{\gamma}{\vartheta} - \frac{\gamma'}{V + \vartheta} + \frac{s}{\vartheta} \tag{A.5.3}$$

$$d_{df}^* = \frac{\tau_{df}^*}{V} = \frac{\gamma}{V} - \frac{\vartheta\gamma'}{V(V + \vartheta)}. \tag{A.5.4}$$

In the TB policy, we can obtain the optimal disposal fee represented by τ_{tb}^*, by solving equations (5.17b), (5.18), and (5.27):

$$\tau_{tb}^* = \gamma - \frac{\vartheta}{V + \vartheta}\gamma' + \frac{V}{V + \vartheta}s. \tag{A.5.5}$$

Substituting (A.5.5) into (5.17b) and (5.18), we can obtain:

$$r_{tb}^* = \frac{\tau_{tb}^*}{\vartheta} = \frac{\gamma}{\vartheta} - \frac{\gamma'}{V + \vartheta} + \frac{V}{\vartheta(V + \vartheta)}s \tag{A.5.6}$$

$$d_{tb}^* = \frac{\tau_{tb}^*}{V} = \frac{\gamma}{V} - \frac{\vartheta\gamma'}{V(V + \vartheta)} + \frac{1}{V + \vartheta}s. \tag{A.5.7}$$

First, we can calculate Δ_1 directly for comparing the DF policy with the TB policy:

$$\Delta_1 = \frac{s}{2(V + \vartheta)}(s + 2\gamma'). \tag{A.5.8}$$

Note that $s + 2\gamma' > 0$, since it is assumed that $s + \gamma > \pi > 0$ and $\gamma' > \gamma$ are satisfied. Therefore, the DF policy is superior to the TB policy if $s > 0$ and vice versa, as shown above.

Now suppose that the price of the collected-for-recycling good 2 is positive. This ensures that the DF policy is superior to the TB policy. Therefore, we should obtain the critical value of the marginal transaction cost, π^*, to compare the DF policy with the DR policy. It can be shown that the sign of $\frac{\partial \pi^*}{\partial s}$ is negative, by using (5.41). We can directly calculate $\frac{\partial \pi^*}{\partial s}$:

$$\frac{\partial \pi^*}{\partial s} = -\frac{\pi}{s + \gamma - \pi} < 0.$$

Conversely, suppose that the price of the collected-for-recycling good 2 is negative. Note that the TB policy is superior to the DF policy. Therefore, the critical value of the marginal transaction cost, π^{**}, should be obtained to compare the DR policy with the TB policy. Using (5.42), we can calculate $\frac{\partial \pi^{**}}{\partial s}$:

$$\frac{\partial \pi^{**}}{\partial s} = 1 + \frac{-\gamma + \frac{\vartheta}{V + \vartheta}\gamma' - \frac{V}{V + \vartheta}s}{s + \gamma - \pi}.$$

Note that $s < 0$. Then, if γ' is sufficiently large relative to γ, that is, $\gamma - \frac{\vartheta}{V + \vartheta}\gamma' < 0$, we can obtain the following inequality:

$$\frac{\partial \pi^{**}}{\partial s} > 1.$$

The critical values of the marginal transaction cost represented by π^* and π^{**} intersect each other on the line of $s = 0$ in the $(\pi - s)$-plane, and the coordinate of this point can be directly calculated:

$$\pi = \gamma - \left(\gamma^2 + \frac{\vartheta}{V}\{(\gamma' - \gamma^2) - \frac{V}{V + \vartheta}\gamma'^2\}\right)^{\frac{1}{2}}.$$

These results are summarized in Figure 5.3.

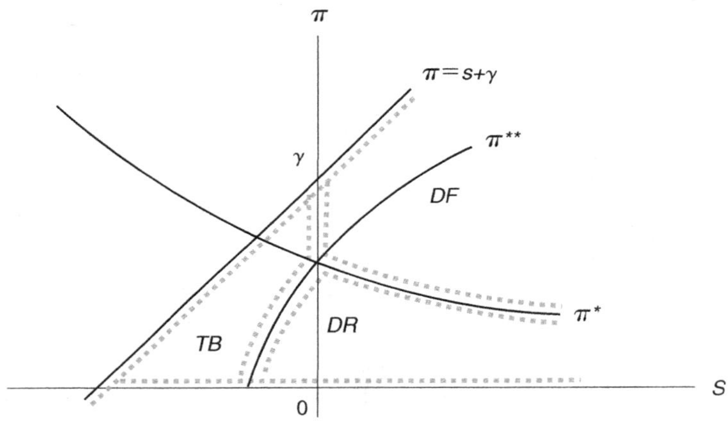

Figure 5.3 An illustration of the second-best policy.

6 The durability of goods and the right recycling policy for household waste

6.1 Introduction[1]

Let us go back to the basic result obtained in Chapter 4: that the social optimum can be attained by either a DF policy or EPR policies (including a DR policy and a TB policy). In this chapter we show that the basic result cannot apply to *durable* goods, even in an economy without transaction costs. It is important to differentiate between *durable* and *nondurable* goods when considering recycling and waste disposal policies.

An advance disposal fee (ADF) policy that is also an EPR policy has been adopted in many countries, especially for the recycling and waste disposal of end-of-life durable goods (see Section 3.4). For the recycling and waste disposal of end-of life home appliances the European Union (EU) decided to introduce an ADF policy (e.g., the Directive on Waste Electric and Electronic Equipment, which came into effect in 2003), as have some Asian countries (notably, Korea and Taiwan). In Japan a DF policy has been adopted for end-of-life home appliances, but Japan, along with some EU countries, has decided to introduce an ADF policy for end-of-life vehicles.

Under an ADF policy consumers are required to pay a recycling and disposal fee for end-of-life goods in advance when they purchase new goods. The advance disposal fees collected from consumers are used for subsidizing recycling and disposal activities. Therefore, an ADF policy can be considered a DR policy, as noted in Section 3.3. In this chapter we consider an ADF policy as a representative EPR policy for recycling end-of-life *durable* goods and we compare it with a DF policy.

In our earlier chapters, consumers' choices were limited to recycling, proper disposal, and illegal disposal. But *durable* goods differ from *nondurable* goods especially in two respects: they are traded in the used-goods market, and consumers who own durable goods can also choose between repairing them or disposing of them when they are out of order. Therefore, a consumer dealing with *durable* goods has a greater variety of options: to buy a new good, to buy or sell a used good, to repair a good if it is out of order, or to dispose of it. The model used in this chapter comprehensively covers all these possible choices as the effects of recycling and waste disposal policies on consumer behavior are studied.

It is shown that the theoretical result of the coincidence between a DF policy and an ADF policy is only applicable to *nondurable* goods. First, we consider an ideal economy where there are no transaction costs. We show that the social optimum can be attained not by an ADF policy but by a DF policy. We also examine the distortion generated under an ADF policy. If an ADF policy is adopted, durable goods are excessively consumed and the lifetime of durable goods becomes shorter than the social optimum.

Our analysis will show that the social optimum can be attained by an alternative policy in an economy without transaction costs. This alternative first-best policy, which can be termed a "revised ADF policy," consists of imposing an advance disposal fee and subsidizing the repair of defective used goods.

Now, all recycling and waste disposal policies incur transaction costs, for a variety of reasons. A DF policy generates a high transaction cost, since monitoring is required to prevent illegal disposal. A revised ADF policy complicates the whole system more than an ADF or a DF policy, because it includes not only a tax but also a subsidy. As stated before, any subsidy generates a high transaction cost. In this respect, an ADF policy has some transaction cost advantages over other policies, though it cannot attain the social optimum.

In Section 6.2, assuming an economy without transaction costs, we derive the competitive equilibria corresponding to DF and ADF policies and compare them with each other, extending the famous durable goods model of Hendel and Lizzeri (1999a, 1999b). In Section 6.3, we derive the first-best policy. In Section 6.4, we examine the conditions required to assert that an ADF policy is the second-best by comparative statics. Section 6.5 summarizes this chapter.

6.2 Competitive equilibrium

Our economy is analyzed in a discrete-time, infinite-horizon model. In this section, we derive the competitive equilibrium in an ideal economy without transaction costs, including monitoring costs to prevent illegal disposal. Note that no used good can be disposed of illegally in such an economy.

We assume that durable goods are produced by a competitive producer and that the production cost per durable good is the constant c. In our model, durable goods can be used at most for two periods.

It is assumed that new goods (in the first period) can never be out of order, but used goods (in the second period) can. The probability that a used good functions normally is represented by q. If a used good is out of order (the probability is $1 - q$), it can be either repaired or disposed of. If it is repaired, it will work well. We assume that defective used goods are repaired by a competitive repair firm. The constant repair cost is denoted by r. Conversely, if the good is disposed of, a constant net disposal cost denoted by d should be paid to a competitive recycling and waste disposal firm.[2] Note that, after a good has been used for two periods, it must be disposed of.

Our competitive economy consists of N consumers. For simplicity, the number of consumers will be normalized and set to one ($N = 1$). Each consumer possesses at most one durable good. Under the normalization assumption, the number of new goods (or used goods) means the proportion of consumers who purchase new goods (or used goods). The willingness to pay (WTP) for a new durable good is represented by w. Each consumer holds a specific value of w and for simplicity w is assumed to be uniformly distributed in a range represented by $[0, \bar{w}]$. Therefore, the density represented by f is constant. This assumption, however, does not have any effect on our results at all. In addition, the WTP for used goods is assumed to be kw ($0<k<1$ is constant).

At the beginning of each period, each consumer takes an option from among the four options stated below, according to his or her valuation. Option 1 represents purchasing a new good and selling it in the used goods market at the beginning of the next period; Option 2 represents purchasing a used good and repairing it if it is out of order; Option 3 represents purchasing a used good and disposing of it if it is out of order; and Option 4 represents purchasing neither a new nor a used good. Each consumer chooses the option that maximizes his or her consumer surplus.

It is assumed that the quality of a used good is uncertain for both buyers and sellers. After purchasing a used good, the buyer finds out about the quality. Hence, before purchasing a used good, a buyer of a used good has to be prepared to decide whether to repair it (Option 2) or dispose of it (Option 3) if it turns out to be unusable.

In Sections 6.2.1 and 6.2.2 below, we analyze the competitive market equilibrium corresponding to an ADF policy and a DF policy, respectively. In Section 6.2.3, we compare them with each other. We will restrict our analysis to the steady-state rational expectations equilibrium.

6.2.1 Competitive market equilibrium with an ADF policy

Assume that an ADF policy is adopted in our competitive economy. Under this policy, producers of durable goods collect an advance disposal fee from purchasers of new goods. The collected advance disposal fee will be allocated to the recycling and waste disposal firm when the durable goods are disposed of.

First, we consider consumer behavior. Each consumer can take the four options stated above. A consumer's surplus changes with each of the four options. The four options and the corresponding consumer surplus evaluated at the beginning of the decision period are denoted by CS and are listed below.

Option 1: Purchasing a new good and selling it at the beginning of the next period.

$$CS^1 = w - P - s + \delta P^u$$

Option 2: Purchasing a used good and repairing it if it is out of order.

$$CS^2 = kw - P^u - (1-q)r$$

Option 3: Purchasing a used good and disposing of it if it is out of order.

$$CS^3 = qkw - P^u$$

Option 4: Purchasing neither a new good nor a used good.

$$CS^4 = 0$$

The price of the new goods and used goods are denoted by P and P^u, respectively. The discount factor is denoted by δ. The advance disposal fee collected from purchasers of new goods is represented by s; it is allocated to the recycling and waste disposal firm.

Figure 6.1 describes how a consumer maximizes his or her CS. At the beginning of period t, consumers with $w \in [w_1(t), \overline{w}]$ take Option 1, consumers with $w \in [w_2(t), w_1(t)]$ take Option 2, consumers with $w \in [w_3(t), w_2(t)]$ take Option 3, and consumers with $w \in [0, w_3(t)]$ take Option 4. Stationarity requires $w_1(t) = w_1(t+1)$, $w_2(t) = w_2(t+1)$, and $w_3(t) = w_3(t+1)$. We represent these stationary values by w_1, w_2, and w_3 respectively.

We can calculate w_1, w_2, and w_3 from the three equations, $CS^1 = CS^2$, $CS^2 = CS^3$, and $CS^3 = CS^4$, respectively:

$$w_1 = \frac{1}{1-k}(P - (1+\delta)P^u - (1-q)r + s) \tag{6.1}$$

$$w_2 = \frac{r}{k} \tag{6.2}$$

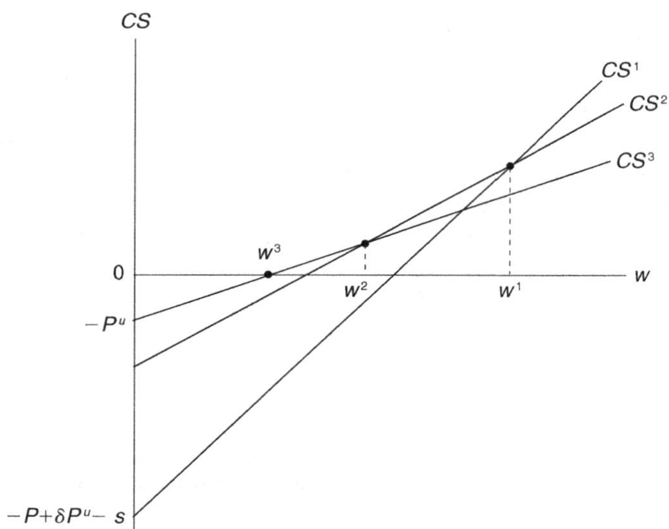

Figure 6.1 Relationship between WTPs for new goods w and consumer surplus corresponding to four options.

$$w_3 = \frac{P^u}{qk}. \tag{6.3}$$

Note that the four options do not include the following option, in which a consumer purchases a new good and keeps it in the next period. It can be easily shown that this option is dominated by other options.[3] A similar result is obtained in Hendel and Lizzeri (1999b).

Because we assume a competitive market, the profit of each firm must be zero at the steady-state equilibrium. The advance disposal fee collected from each purchaser of new goods denoted by s can be calculated under the condition that the expected profit of the recycling and waste disposal firm must be zero:

$$s = \frac{\left[\delta^2 d \int_{w_2}^{w_1} dF + \delta^2 qd \int_{w_3}^{w_2} dF + \delta(1-q)d \int_{w_3}^{w_2} dF\right]}{\int_{w_1}^{\bar{w}} dF}. \tag{6.4}$$ [4]

Note that s is equivalent to the expected average net disposal cost evaluated in the period when a new good is purchased. The profit of a producer in each period at the steady state, Π^{ADF}, is represented by

$$\Pi^{ADF} = (P-c)\int_{w_1}^{\bar{w}} dF.$$

Therefore, competitive equilibrium requires

$$(P-c)\int_{w_1}^{\bar{w}} dF = 0. \tag{6.5}$$

Simultaneously, the following equation representing the market clearance condition of the used-goods market must be satisfied:

$$\int_{w_1}^{\bar{w}} dF dF - \int_{w_3}^{w_1} dF = 0. \tag{6.6}$$

We can determine the steady-state equilibrium in the competitive market from equations (6.1) to (6.6). For further analysis we use (6.1), (6.3), and (6.4) and rewrite (6.5) as

$$((1-k)w_1 + (1+\delta)qkw_3 + (1-q)r - c)\int_{w_1}^{\bar{w}} dF - \delta^2 d\int_{w_2}^{w_1} dF$$
$$-\delta^2 qd\int_{w_3}^{w_2} dF - \delta(1-q)d\int_{w_3}^{w_2} dF = 0. \tag{6.7}$$

Note that the left-hand side of this equation represents the steady-state producer's profit, Π^{ADF}. By solving w_3 as a function of w_1 from (6.6) and substituting it into

(6.7), we can easily see that Π^{ADF} is a concave function of w_1. Since w is assumed to be uniformly distributed, Π^{ADF} is a quadratic function of w_1. Therefore, there are at most two steady-state solutions of w_1. The larger one, however, cannot be a competitive equilibrium. To understand the reason for this, let us suppose that the larger solution is a competitive equilibrium. In this equilibrium, a producer of durable goods can receive a positive profit by increasing production. This is a contradiction. It follows that the competitive equilibrium is unique if it exists.

6.2.2 Competitive market equilibrium with a DF policy

Alternatively, we adopt a DF policy and derive the corresponding steady-state equilibrium. Under a DF policy, consumers are required to pay the net disposal cost when they dispose of durable goods. Note that we ignore the possibility of illegal disposal, since we assume that the cost of monitoring can be neglected.

Under a DF policy, a consumer's surplus corresponding to the four options can be calculated as follows:

$$CS^1 = w - P + \delta P^u$$

$$CS^2 = kw - P^u - (1-q)r - \delta d$$

$$CS^3 = qkw - P^u - \delta qd - (1-q)d$$

$$CS^4 = 0.$$

Similarly, we can calculate w_1, w_2, and w_3.

$$w_1 = \frac{1}{1-k}(P - (1+\delta)P^u - (1-q)r - \delta d) \tag{6.8}$$

$$w_2 = \frac{r - (1-\delta)d}{k} \tag{6.9}$$

$$w_3 = \frac{P^u + \delta qd + (1-q)d}{qk}. \tag{6.10}$$

The steady-state producer's profit, Π^{DF}, can also be rewritten as

$$\Pi^{DF} = (P-c)\int_{w_1}^{\bar{w}} dF.$$

The competitive equilibrium requires that the steady-state producer's profit be zero. That is, the following equation must be satisfied:

$$\begin{aligned}((1-k)w_1 &+ (1+\delta)(qkw_3 - \delta qd - (1-q)d) \\ &+ (1-q)r + \delta d - c)\int_{w_1}^{\bar{w}} dF = 0.\end{aligned} \tag{6.11}$$

Note that the profit of the recycling and waste disposal firm is always zero in a DF policy. In addition, the market clearance condition for the used goods, (6.6) in the previous section, must be satisfied. The left-hand side of (6.11) represents the steady-state producer's profit denoted by Π^{DF}. By using (6.6), we easily see that Π^{DF} is also a concave quadratic function of w_1. In this case, there are two solutions and the larger one is $\bar{w}$. This solution cannot be a competitive equilibrium for the same reason stated in the previous section. It follows that the competitive equilibrium is unique.

6.2.3 Comparison between a DF policy and an ADF policy

In this section, we compare two equilibria. One is the steady-state equilibrium if an ADF policy is adopted, and the other is the steady-state equilibrium corresponding to a DF policy. To compare them, we rewrite the steady-state producer's profits, Π^{ADF} and Π^{DF}, obtained in the previous sections, as follows:

$$\Pi^{ADF} = ((1-k)w_1 + (1+\delta)qkw_3 + (1-q)r - c)\int_{w_1}^{\bar{w}} dF$$

$$- \delta^2 d\int_{w_2}^{w_1} dF - \delta^2 qd\int_{w_3}^{w_2} dF - \delta(1-q)d\int_{w_3}^{w_2} dF$$

$$\Pi^{DF} = ((1-k)w_1 + (1+\delta)qkw_3 + (1-q)r - c)\int_{w_1}^{\bar{w}} dF - (1-q+\delta^2 q)d\int_{w_1}^{\bar{w}} dF.$$

We represent the solution of (w_1, w_2, w_3) in an ADF policy and a DF policy by (w_1^A, w_2^A, w_3^A) and (w_1^D, w_2^D, w_3^D) respectively. Note that w_2^A and w_2^D are given by (6.2) and (6.9) respectively. In the competitive equilibrium, the steady-state producer's profit must be zero, namely, $\Pi^{ADF}(w_1^A, w_2^A, w_3^A) = 0$ and $\Pi^{DF}(w_1^D, w_2^D, w_3^D) = 0$. By solving w_3 as a function of w_1 from (6.6) and substituting it into these equilibrium conditions, the equilibrium producer's profits, Π^{ADF} and Π^{DF} are concave functions of w_1^A and w_1^D respectively. As we saw in the previous sections, there exist at most two equilibrium candidates and only the smaller one can be the equilibrium. The following proposition compares the competitive equilibria in the two policies to each other.

Proposition 6.1. *Consider an economy without transaction costs. The competitive equilibria represented by (w_1^A, w_2^A, w_3^A) in an ADF policy and (w_1^D, w_2^D, w_3^D) in a DF policy satisfy the following inequalities: $w_3^A < w_3^D < w_2^D < w_2^A < w_1^A < w_1^D$.*

Proof. We can easily see that w_2^A is strictly larger than w_2^D from (6.2) and (6.9); that is

$$w_2^D < w_2^A. \tag{6.12}$$

Now suppose that $w_1^A = w_1^D$. Then we obtain $w_3^A = w_3^D$ from (6.6). We can see $\Pi^{ADF} - \Pi^{DF}$ is positive as follows:

$$
\begin{aligned}
&\Pi^{ADF}(w_1^D, w_2^A, w_3^D) - \Pi^{DF}(w_1^D, w_2^D, w_3^D) \\
&= -\delta^2 d \int_{w_2^A}^{w_1^D} dF - \delta^2 qd \int_{w_3^D}^{w_2^A} dF - \delta(1-q)d \int_{w_3^D}^{w_2^A} dF \\
&\quad + (1 - q + \delta^2 q)d\left(\int_{w_2^A}^{w_1^D} dF + \int_{w_3^D}^{w_2^A} dF \right) \\
&= (1-\delta^2)(1-q)d \int_{w_2^A}^{w_1^D} dF + (1-\delta)(1-q)d \int_{w_3^D}^{w_2^A} dF \\
&= (1-\delta)(1-q)d\left((1+\delta) \int_{w_2^A}^{w_1^D} dF + \int_{w_3^D}^{w_2^A} dF \right) \\
&= (1-\delta)(1-q)d\left(\int_{w_3^D}^{w_1^D} dF + \delta \int_{w_2^A}^{w_1^D} dF \right) \\
&> (1-\delta)(1-q)d\left(\int_{w_3^D}^{w_1^D} dF - \delta \int_{w_1^D}^{\bar{w}} dF \right) \\
&> 0. \qquad (6.13)
\end{aligned}
$$

The first equality and the last inequality follow from equation (6.6). From inequality (6.13) we can conclude that $w_1^A < w_1^D$ and $w_3^A < w_3^D$. Putting together all these results, we obtain the desired inequalities.

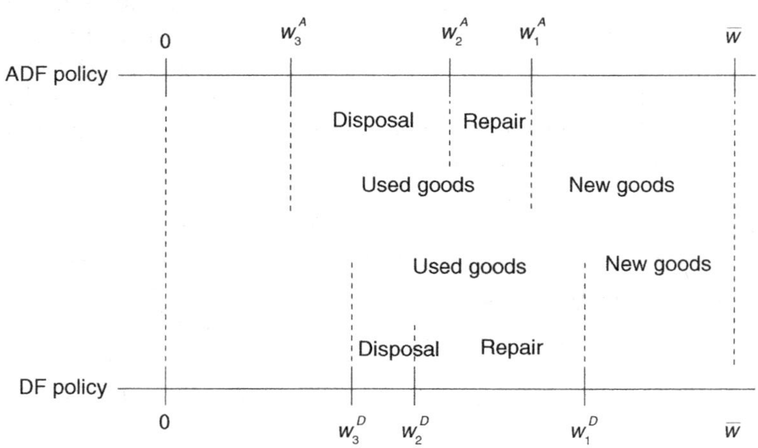

Figure 6.2 Comparisons of competitive market equilibriums in ADF policy and DF policy.

This proposition means that an ADF policy results in excessive consumption and disposal compared with a DF policy as described in Figure 6.2.

Assume that used goods work without any repair ($q = 1$). In this special case, it is not rational to choose Option 3 (purchasing a used good and disposing of it if it is out of order). Equation (6.13) shows that the equilibrium solutions in the two policies coincide with each other if $q = 1$.

Corollary 6.1. *If used goods work without any repair ($q = 1$), an ADF policy and a DF policy will produce the same result.*

If used goods are out of order ($q < 1$), however, consumer choices in an ADF policy will differ from those in a DF policy ($w_2^A > w_2^D$). This difference is because a consumer is exempted from the disposal fee in an ADF policy, and this net disposal fee exemption distorts a consumer's choice toward repair. Therefore, in an ADF policy, Option 3 (disposing of a defective used good) becomes more attractive. This consideration tempts some consumers who might choose Option 2 and Option 4 into choosing Option 3, which means that w_3 decreases—i.e., the used good market expands. Note that w_1 also decreases (i.e., the new good market expands) if w_3 decreases (the used good market expands). This is an economic interpretation of Proposition 6.1.

Now we define the rate of repair as an index of durability:

Rate of repair = the proportion of repaired used goods to defective used goods

$$= \frac{\int_{w_2}^{w_1} dF}{(1-q)\int_{w_3}^{w_1} dF} = \frac{\int_{w_2}^{w_1} dF}{(1-q)\int_{w_1}^{\bar{w}} dF}.$$

We can easily see from Proposition 6.1 that the lifetime is shorter in an ADF policy than in a DF policy.

Corollary 6.2. *The rate of repair (defined above) is lower in an ADF policy than in a DF policy:*

$$\frac{\int_{w_2^A}^{w_1^A} dF}{(1-q)\int_{w_3^A}^{w_1^A} dF} < \frac{\int_{w_2^D}^{w_1^D} dF}{(1-q)\int_{w_3^D}^{w_1^D} dF}.$$

6.3 The first-best policy

In this section we still consider an economy without transaction costs, and derive the first-best policy. First of all, we derive the social optimal solution. The social optimum problem can be formalized as follows:

$$\underset{w_1(t),\, w_2(t),\, w_3(t)}{Max} \quad SW = \sum_{t=0}^{\infty} \delta^t \left[\begin{array}{l} \int_{w_1(t)}^{\bar{w}} w dF + \delta \int_{w_2(t+1)}^{w_1(t+1)} kw dF + \delta q \int_{w_3(t+1)}^{w_2(t+1)} kw dF \\ - c \int_{w_1(t)}^{\bar{w}} dF - \delta(1-q)\, r \int_{w_2(t+1)}^{w_1(t+1)} dF - \delta^2 d \int_{w_2(t+1)}^{w_1(t+1)} dF \\ - \delta(1-q) d \int_{w_3(t+1)}^{w_2(t+1)} dF - \delta^2 qd \int_{w_3(t+1)}^{w_2(t+1)} dF \end{array} \right]$$

subject to

$$\int_{w_1(t)}^{\bar{w}} dF - \int_{w_3(t+1)}^{w_1(t+1)} dF = 0. \tag{6.14}$$

We define $w_1(t)$, $w_2(t)$, and $w_3(t)$, respectively, as:

$w_1(t)$: the minimum WTP of new-good owners in period t

$w_2(t)$: the minimum WTP of used-good owners whose used goods will be repaired if they are out of order in period t

$w_3(t)$: the minimum WTP of used-good owners whose used goods will be disposed of if they are out of order in period t

Note that the objective function itself assumes $w_1(t) \geq w_2(t) \geq w_3(t)$ implicitly.[5] The first three terms in the bracket of the objective function represent consumer surplus in period t. The fourth term and the fifth term represent production costs and repair costs, respectively. The residuals represent net disposal costs accrued in period t. Then the constraint (6.14) means that a new good in period t becomes a used good in period t+1. For simplicity, we restrict our analysis to the inner solution. Equivalently, we assume $w_1(t) > w_2(t) > w_3(t)$.

Define Lagrange function Γ as

$$\Gamma = SW + \sum_{t=0}^{\infty} \lambda(t) \left(\int_{w_1(t)}^{\bar{w}} dF - \int_{w_3(t+1)}^{w_1(t+1)} dF \right).$$

Then we obtain the first-order conditions:

$$\frac{\partial \Gamma}{\partial w_1(t)} = \{\delta^t(-w_1(t) + c) - \lambda(t) + \delta^{t-1}(\delta k w_1(t) - \delta(1-q)r - \delta^2 d) - \lambda(t-1)\} f = 0 \tag{6.15}$$

$$\frac{\partial \Gamma}{\partial w_2(t)} = \{\delta^{t-1}(-\delta k w_2(t)) + q\delta k w_2(t) + \delta(1-q)r + \delta^2 d$$

$$- \delta(1-q)d - \delta^2 qd)\} f = 0 \tag{6.16}$$

$$\frac{\partial \Gamma}{\partial w_3(t)} = \{\delta^{t-1}(-\delta qkw_3(t) + \delta(1-q)d + \delta^2 qd) + \lambda(t-1)\} f = 0. \tag{6.17}$$

Noting that $\frac{\lambda(t-1)}{\delta^{t-1}}$ is constant in the steady-state solution, we can obtain the following equation by arranging the three equations (6.15), (6.16), and (6.17):

$$(1-k)\,w_1 + (1+\delta)(qkw_3 - \delta qd - (1-q)d) + (1-q)r + \delta d - c = 0. \tag{6.18}$$

The constraint (6.14) can be rewritten at the steady state as follows:

$$\int_{w_1}^{\bar{w}} dF - \int_{w_3}^{w_1} dF = 0. \tag{6.19}$$

We can obtain the steady-state solution w_1 and w_3 from (6.18) and (6.19). We can easily solve equation (6.16) and derive w_2:

$$w_2 = \frac{r-(1-\delta)d}{k}. \tag{6.20}$$

If equations (6.18), (6.19), and (6.20) are compared with (6.11), (6.6), and (6.9), respectively, we can see that the competitive market with a DF policy is efficient. This result is summarized in the next proposition.

Proposition 6.2. *Assuming an economy without transaction costs, the competitive market will produce the social optimum outcome if a DF policy is adopted.*

It follows from Proposition 6.1 and Proposition 6.2 that the competitive market under an ADF policy is inefficient:

Corollary 6.3. *If an ADF policy is adopted, the social optimum cannot be attained.*

Let us consider the reason why the competitive market under an ADF policy is inefficient. In a DF policy, the net disposal fee gives each disposer a strong incentive to repair defective used goods. In other words, Option 2 (repairing a defective used good) becomes a more attractive option relative to Option 3 (disposing of a defective used good). Alternatively, suppose that an ADF policy is adopted. Under this policy, purchasers of new goods are required to pay the advance disposal fee denoted by s. As a result, the price of used goods denoted by P^u increases. The higher prices of used goods, however, do not give a consumer sufficient incentive to choose Option 2 instead of Option 3. Note that consumers choosing Option 2 and Option 3 are purchasers of used goods. The increase of the used-goods price works with Option 2 and Option 3 in a similar way. In short, an ADF policy fails to give purchasers of used goods sufficient incentive to repair used goods when they are out of order.

As seen in Proposition 6.2, a DF policy is the first-best policy. However, the first-best outcomes can be obtained by an alternative policy, one that includes a subsidy for repair. Note that we need to strengthen the incentive for consumers to repair defective used goods, if a disposer is exempted from payment of the net disposal cost.

In this alternative policy, a purchaser of a new good is taxed by an amount of x, represented by $x = \delta(1-q)d + \delta^2 qd$. Simultaneously, a consumer who repairs a defective used good is given a subsidy y that is set at $y = (1-\delta)d$. This alternative policy can be considered an amended version of an ADF policy. We can call it a revised ADF policy. The following proposition states that the revised ADF policy can also attain the social optimum.

Proposition 6.3. *We can also attain the optimal outcome by a revision of ADF policy. This revised ADF policy consists of a tax on the purchase of new goods and a subsidy for the repair of used goods. The tax per new good is denoted by* x *and the subsidy for repair is denoted by* y*. They are set to* $x = \delta(1-q)d + \delta^2 qd$ *and* $y = (1-\delta)d$*, respectively.*

Proof. If the revised ADF policy is adopted, a consumer's surplus corresponding to each of the four options can be calculated as follows:

$$CS^1 = w - P + \delta P^u - x = w - P + \delta P^u - \delta(1-q)d - \delta^2 qd$$

$$CS^2 = kw - P^u - (1-q)r + (1-q)y = kw - P^u - (1-q)r + (1-q)(1-\delta)d$$

$$CS^3 = qkw - P^u$$

$$CS^4 = 0.$$

We can easily show that the equilibrium conditions (6.9) and (6.11) obtained in Section 6.2.2 must be satisfied in this policy. We obtain the desired result.

6.4 Transaction cost and ADF policy as the second-best policy

We obtained two first-best policies, namely, a DF policy and a revised ADF policy. However, all recycling and waste disposal policies generate associated transaction costs. Taking transaction costs into consideration, we can never attain the social optimum. Suppose that we adopt a DF policy. Consumers are then tempted to dispose of obsolete or defective used goods illegally. Significant monitoring costs are needed to prevent illegal disposal. Alternatively, let us suppose that a revised ADF policy is adopted. Since a revised ADF policy includes a subsidy for the repair of used goods, the subsidy will also generate significant transaction costs.[6]

On the other hand, an ADF policy also generates a transaction cost, because an ADF policy includes a recycling subsidy. However, it is recycling and waste disposal firms rather than consumers that are given recycling subsidies.[7] Since the number of recycling and waste disposal firms is smaller than the number of consumers, it follows that a transaction cost associated with an ADF policy is not as large as expected.

For these reasons, we can reasonably expect that an ADF policy has some transaction cost advantages over a DF policy and a revised ADF policy, though it cannot attain the first-best outcome. Hence, the benefit of the latter turns on the relative size of its transaction cost advantage and the welfare loss associated with the inefficiency of an ADF policy.

Under some circumstances an ADF policy may therefore be second-best. The next proposition implies that an ADF policy can be the second-best policy either if net disposal cost is relatively low or if repair cost is relatively high.

Proposition 6.4. *Denote by SW^F the maximum social welfare associated with the first-best outcome (analyzed in the preceding section). On the other hand, denote by SW^A the social welfare realized by an ADF policy (analyzed in Section 6.2.1). Then the social welfare loss ($SW^F - SW^A$) is an increasing function of net disposal cost d. In addition, it is a decreasing function of repair cost r.*

Proof. See Appendix 6.

6.5 Summary of this chapter

In recent years a DF policy has been replaced by an ADF policy for recycling and waste disposal of end-of-life *durable* goods. We show that the theoretical result of the coincidence between a DF policy and an ADF policy obtained in the existing literature is only applicable to *nondurable* goods. Our results shed light on the importance of differentiating *durable* and *nondurable* goods when it comes to a recycling and waste disposal policy. It is shown that the competitive market with an ADF policy is inefficient and a DF policy is the optimal policy. The reason for this result is that an ADF policy fails to give owners of used goods sufficient incentive to repair defective used goods.

If an ADF policy is adopted, the result is greater consumption and disposal than is consistent with the social optimum. In addition, under an ADF policy the lifetime of a durable good is shorter than what is consistent with the social optimum. The optimal outcome can also be attained by revising an ADF policy. A revised ADF policy subsidizes the repair of defective used goods.

When we take transaction costs into consideration, however, the social optimum becomes unattainable. In this case, an ADF policy may have some transaction cost advantages over other policies, even though this policy cannot attain the social optimum. We show that an ADF policy can be second-best either if the net disposal cost is relatively low or if the repair cost is relatively high.

Appendix

Proof of Proposition 6.4. We denote by SW^F the maximum social welfare associated with the optimal outcome (analyzed in Section 6.3). The superscript F denotes the optimal policy. On the other hand, we will denote by SW^A the social welfare realized by an ADF policy. Note that we can represent SW^A as the maximum value of the objective function in the following problem (call it PA): (PA)

$$\underset{w1(t),\, w2(t),\, w3(t)}{Max} \quad SW = \sum_{t=0}^{\infty} \delta^t \left[\begin{array}{l} \int_{w_1(t)}^{\bar{w}} wdF + \delta \int_{w_2(t+1)}^{w_1(t+1)} kwdF + \delta q \int_{w_3(t+1)}^{w_2(t+1)} kwdF - c \int_{w_1(t)}^{\bar{w}} dF \\ - \delta(1-q) r \int_{w_2(t+1)}^{w_1(t+1)} dF \\ - \left(\hat{s}(w^A{}_1, w^A{}_2, w^A{}_3) / \int_{w_1^A}^{\bar{w}} dF \right) d \int_{w_1(t)}^{\bar{w}} dF \end{array} \right]$$

subject to (6.14), where $\hat{s}(w_1, w_2, w_3) = \delta^2 \int_{w_2}^{w_1} dF + \delta^2 q \int_{w_3}^{w_2} dF + \delta(1-q) \int_{w_3}^{w_2} dF.$

Note that we substitute $(w_1{}^A, w_2{}^A, w_3{}^A)$ obtained in Section 6.2.1 into $\hat{s}(w_1, w_2, w_3) / \int_{w_1}^{\bar{w}} dF$ before optimization.

A.6.1 Comparative statics with respect to net disposal cost d

First, we examine whether the social welfare loss associated with an ADF policy increases or decreases if net disposal cost d increases. Applying the envelope theorem to the maximization problem formulated in Section 6.3, we obtain

$$\frac{\partial SW^F}{\partial d} = -\sum_{t=0}^{\infty} \delta^t \left(\delta^2 \int_{w_2(t+1)}^{w_1(t+1)} dF + \delta^2 q \int_{w_3(t+1)}^{w_2(t+1)} dF + \delta(1-q) \int_{w_3(t+1)}^{w_2(t+1)} dF \right).$$

In addition, evaluating it at the steady state, we can see

$$\frac{\partial SW^F}{\partial d} = -\Delta \hat{s}(w_1^D, w_2^D, w_3^D) < 0, \tag{A.6.1}$$

where $\Delta = \sum_{t=0}^{\infty} \delta^t$. Note that a DF policy can attain the optimal outcome in an economy without transaction costs.

Again, by applying the envelope theorem to (PA) we can obtain $\frac{\partial SW^A}{\partial d}$. The following result is evaluated at the steady state.

$$\frac{\partial SW^A}{\partial d} = -\Delta \hat{s}(w_1^A, w_2^A, w_3^A) < 0. \tag{A.6.2}$$

First, note that $SW^F - SW^A = 0$, if $d=0$. Second, to inspect the effect of d on $SW^F - SW^A$, we differentiate it with respect to d. By applying the envelope theorem, we can obtain the following result:

$$\begin{aligned}
\frac{1}{\Delta}\frac{\partial(SW^F - SW^A)}{\partial d} &= \hat{s}(w_1^A, w_2^A, w_3^A) - \hat{s}(w_1^D, w_2^D, w_3^D) \\
&= \delta^2 \int_{w_2^A}^{w_1^A} dF + (\delta^2 q + \delta(1-q)) \int_{w_3^A}^{w_2^A} dF - \delta^2 \int_{w_2^D}^{w_1^D} dF - (\delta^2 q + \delta(1-q)) \int_{w_3^D}^{w_2^D} dF \\
&= -\delta^2 \left(\int_{w_1^A}^{w_1^D} dF + \int_{w_2^D}^{w_2^A} dF \right) + (\delta^2 q + \delta(1-q)) \left(\int_{w_2^D}^{w_2^A} dF + \int_{w_3^A}^{w_3^D} dF \right) \\
&= -\delta^2 \int_{w_1^A}^{w_1^D} dF + \delta(1-\delta)(1-q) \int_{w_2^D}^{w_2^A} dF + (\delta^2 q + \delta(1-q)) \int_{w_3^A}^{w_3^D} dF \\
&= \delta\{2(1-\delta)(1-q) + \delta\} \int_{w_1^A}^{w_1^D} dF + \delta(1-\delta)(1-q) \int_{w_2^D}^{w_2^A} dF > 0.
\end{aligned} \tag{A.6.3}$$

For the derivation of the third equality we use the inequalities represented by $w_3^A < w_3^D < w_2^D < w_2^A < w_1^A < w_1^D$ (Proposition 6.1). The last equality in (A.6.3) follows from the next result (A.6.4). The market clearance for used goods requires that the following equations must be satisfied: $\int_{w_1^D}^{\bar{w}} dF - \int_{w_3^D}^{w_1^D} dF = 0$ and $\int_{w_1^A}^{\bar{w}} dF - \int_{w_3^A}^{w_1^A} dF = 0$ in a DF policy and in an ADF policy, respectively.

Using these equations, we can obtain

$$\int_{w_1^A}^{w_1^D} dF = \int_{w_1^A}^{\bar{w}} dF - \int_{w_1^D}^{\bar{w}} dF = \int_{w_3^A}^{w_1^A} dF - \int_{w_3^D}^{w_1^D} dF = \int_{w_3^A}^{w_3^D} dF - \int_{w_1^A}^{w_1^D} dF.$$

These equations imply

$$\int_{w_3^A}^{w_3^D} dF = 2 \int_{w_1^A}^{w_1^D} dF. \tag{A.6.4}$$

Therefore, we can conclude that the social welfare loss represented by $SW^F - SW^A$ is an increasing function of net disposal cost d.

A.6.2 Comparative statics with respect to repair cost r

Next, we examine whether the social welfare loss increases or decreases if repair cost increases. Applying the envelope theorem to the maximization problem in Section 6.3 and (PA), we obtain $\frac{\partial SW^F}{\partial r}$ and $\frac{\partial SW^A}{\partial r}$ respectively:

$$\frac{\partial SW^F}{\partial r} = -\delta\Delta(1-q)\int_{w_2^D}^{w_1^D} dF < 0$$

$$\frac{\partial SW^A}{\partial r} = -\delta\Delta(1-q)\int_{w_2^A}^{w_1^A} dF < 0.$$

Note that they are evaluated at the steady state. By Proposition 6.1, we can immediately obtain the following inequality:

$$\frac{\partial SW^F}{\partial r} < \frac{\partial SW^A}{\partial r}.$$

Therefore, we can assert that the social welfare loss is a decreasing function of repair cost r.

7 International trade of waste and environmental pollution

7.1 Introduction

Even though municipalities have reduced the volume of waste finally disposed of by compressing, incinerating, and recycling it, the total amount of waste needing final disposal has been on the increase. In several OECD countries, the depletion of final disposal sites has become a serious problem due to opposition to new landfills or incinerators (the so-called NIMBY problem), which has raised disposal costs, especially for hazardous waste. To mitigate this problem, waste has been exported to developing countries. In the 1970s and 1980s in particular, hazardous waste was exported for final disposal from developed countries to developing countries, a process that caused serious environmental pollution.

For years now, global economic policy and its link with dumping rich countries' poisons on their poorer neighbors have been discussed. Lawrence H. Summers wrote an internal memo on December 12, 1991 stating that the World Bank should encourage more migration of the dirty industries to the less developed countries. Three reasons were put forward to support this argument. First, the dumping of a load of toxic waste in a country with the lowest wages is impeccable considering the costs of health-impairing pollution. The measurement of cost depends on the foregone earnings from increased morbidity and mortality. Therefore, activities that produce health-impairing pollution need to be provided in the country with the lowest cost. Second, the costs of pollution are likely to be nonlinear, as the initial increments of pollution have a very low cost. The cost in a less developed country such as some of those in Africa is very low, and it can be increased. Third, income elasticity of demand is expected to be high for a clean environment and health.

However, the solution of exporting waste was no longer an easy option after the Basel Convention (full title: the Basel Convention on the Control of Transboundary Movements of Hazardous Wastes and Their Disposal) was put into effect in 1992. The Basel Convention provided a strong motivation to developed countries to put in place a comprehensive recycling scheme based on extended producer responsibility (EPR).

Although such recycling schemes were intended to recycle end-of-life goods domestically, domestic recycling was confronted with high wage rates. Recycling industries are labor intensive, because material recycling requires complete separation of each type of material making up the end-of-life goods. In addition, there is a huge demand for second-hand goods in developing countries. So it was obviously efficient to export end-of-life goods as second-hand goods or to export them as scrap to be recycled in developing countries after roughly sorting goods by machine into different types of material and removing any hazardous material. Note that the sorting and shipping costs for scrap are paid for by producers or consumers under domestic recycling systems. This stimulates not only domestic recycling but also the exportation of end-of-life goods for reuse or recycling in developing countries. Indeed, not a few recycling systems in OECD countries allow the exportation of scrap to be recycled overseas (see Section 3.5).

The trade in waste has undergone enormous changes as a consequence of the establishment of recycling systems in developed countries that followed the Basel Convention. There has been a decrease in the export of waste for final disposal from developed to developing countries. On the other hand, the export of second-hand products for reuse and scrap for recycling has been increasing.

We focus on end-of-life durable goods such as home appliances and personal computers. They have been called E-waste. As a rule, E-waste, or waste electrical and electronic equipment (WEEE) includes second-hand electrical and electronic equipment (EEE). In this study we distinguish between E-waste scrap and second-hand EEE.

Home appliances and computers used in the US, the EU, and Japan have to a large extent been exported into developing countries. Roughly speaking, it is thought that the global picture of E-waste material flow can be split into two closed regions (Schwarzer *et al.*, 2005; Widmer *et al.*, 2005; and Shinkuma and Huong, 2009). The first region includes the western half of the US, Japan, Korea, China, Hong Kong, and Southeast Asian countries. In this region, the US, Japan, Korea, and a few Southeast Asian countries like Malaysia and Singapore are exporters of end-of-life goods, and the importing countries include China, Vietnam, the Philippines, Indonesia, Cambodia, and Myanmar. However, China recycles almost all the scrap accrued in this region, while Southeast Asian countries export scrap to China after they have reused second-hand goods originally imported from developed countries.

The second region includes the EU, the eastern half of the US, Africa, the Middle East, and South Asia (including India, Pakistan, Bangladesh, and Nepal). In this region, while the US and the EU are the chief exporters of second-hand EEE and E-waste scrap, the other countries are mainly importers of second-hand EEE, with the exception of India and Pakistan, which also import E-waste scrap from other countries. Therefore, in this region India and Pakistan are centers of the recycling activities.

Circulation of resources (including the reuse and recycling of material) around the whole of the world is essential in terms of effective utilization of resources. However, in the centers of E-waste recycling—China, India, and

Pakistan—serious environmental pollution has been generated by improper recycling methods, according to a report published by Basel Action Network and Silicon Valley Toxics Coalition (2002).[1]

As a solution to this problem, several countries have banned the import of second-hand items and E-waste scrap. There is also support for the idea of amending the Basel Convention so that the export from developed to developing countries of scrap that includes any hazardous materials—even for recycling—should be banned. The policy issues are discussed in later chapters.

The primary objective of this chapter is to overview the environmental pollution accrued by recycling E-waste scrap that has been observed in China, India, and Pakistan. A second objective is to present the background to the global picture of E-waste material flows described above.

7.2 Environmental pollution in the recycling centers of E-waste

7.2.1 The definition of E-waste and its hazardousness

E-waste encompasses a broad range of electrical and electronic devices ranging from large home appliances such as refrigerators to air conditioners, personal stereos, hand-held cellular phones, and computers. E-waste contains over a thousand different substances, many of which are toxic, and creates serious pollution upon disposal. Table 7.1 shows materials found in a personal desktop computer.

7.2.2 Metal recovery method applied in China and India

As shown in Table 7.1, the metal recovery process generates most of the revenue from E-waste recycling. PCBs (printed circuit boards) are the most valuable because they include two precious metals, gold and copper. Unfortunately, it is also true that the process of recovering the metals from PCBs seriously pollutes the environment. In this section, we focus on PCB recycling and describe briefly, on the basis of what is said in Basel Action Network and Silicon Valley Toxics Coalition (2002), the metal recovery methods applied in the intensive E-waste recycling centers: Guiyu in China, Karachi in Pakistan, and Delhi in India. The methods applied for metal recovery in the three regions are very similar, so we can concentrate on what is done in Guiyu.

Guiyu in Guangdong Province[2]

Since 1995 Guiyu has been transformed from a poor, rural, rice-growing village to a huge center for recycling E-waste scrap, especially PCBs. Over 75 percent of the 300 individual workshops have been involved in E-waste recycling, with about 100,000 migrant laborers from other provinces. These migrant laborers are employed at a low wage rate ($1.50 per day in 2002).

Table 7.1 Composition of a personal desktop computer

Name	*Content (percentage of total weight)*	*Weight of material (lbs)*	*Value of material*
Plastics	22.9907	13.8	$8.28
Lead	6.2988	3.8	$3.01
Aluminum	14.1723	8.5	$7.46
Germanium	0.0016	<0.1	
Gallium	0.0013	<0.1	
Iron	20.4712	12.3	$1.85
Tin	1.0078	0.6	
Copper	6.9287	4.2	$12.45
Barium	0.0315	<0.1	
Nickel	0.8503	0.51	
Zinc	2.2046	1.32	
Tantalum	0.0157	<0.1	
Indium	0.0016	<0.1	
Vanadium	0.0002	<0.1	
Terbium	<0	<0	
Beryllium	0.0157	<0.1	
Gold	0.0016	<0.1	$18.23
Europium	0.0002	<0.1	
Titanium	0.0157	<0.1	
Ruthenium	0.0016	<0.1	
Cobalt	0.0157	<0.1	
Palladium	0.0003	<0.1	
Manganese	0.0315	<0.1	
Silver	0.0189	<0.1	$3.36
Antimony	0.0094	<0.1	
Bismuth	0.0063	<0.1	
Chromium	0.0063	<0.1	
Cadmium	0.0094	<0.1	
Selenium	0.0016	<0.1	
Niobium	0.0002	<0.1	
Yttrium	0.0002	<0.1	
Rhodium	<0	<0	

Notes

This table is based on Annex I in Basel Action Network and Silicon Valley Toxics Coalition (2002); we have added the value of the material to the original table.

In the calculation of the last column, the current price (as of July 7, 2010) for each material is used: gold is $1,186.13/oz ($18,978.08/lb); silver is $18.533/oz ($296.528/lb); copper is $2.9635/lb; aluminum is $0.8773/lb; lead is $0.7919/lb; scrap iron (Bushelings Mill Pittsburgh FOB delivered) is $0.15/lb; scrap plastics (ABS) are $0.60/lb.

The first step in recycling PCBs involves a de-soldering process. Many hundreds of workers place the PCBs on iron plates that are heated over cans filled with burning coal. On each plate is a pool of molten lead-tin solder. The PCBs are heated until the chips on the surface of the circuit boards are removable. The loosened chips are then sorted between those valuable for resale and those to be sent to the acid chemical strippers for gold recovery. The solder is also collected and later melted off and sold. The chips removed from the boards that cannot be resold go to the acid recovery operation. An aqua regia solution (a mixture of 25 percent pure nitric acid and 75 percent pure hydrochloric acid) is first heated over small fires and poured into plastic tubs full of computer chips. A chemical is then added that precipitates the gold, making it settle to the bottom of the tub.

Once the chips have been removed from the circuit boards, the boards themselves are burned outdoors to retrieve the last remaining metal, copper. The ashes are floated in water to remove lighter ash.

These metal recovery processes can seriously damage the environment and human health. While fans are sometimes used to blow the toxic lead-tin solder fumes away, exposure on a daily basis is likely to be very damaging to human beings. The open burning process is bound to emit substantial quantities of harmful heavy metals, dioxins, beryllium, and polycyclic aromatic hydrocarbons. The waste acid is dumped directly into the river or onto the ground without any proper treatment.

The environmental pollution generated by the metal recovery processes is considerable, and its severity has been reported in recent academic papers. Here we focus our attention on those studies that deal with the environmental pollution at Guiyu. A couple of studies that investigated the effects of E-waste recycling on the environment around Guiyu were reviewed by Wong *et al.* (2007).

According to this review, the concentrations of persistent organic pollutants (POPs) (such as flame retardants, dioxins/furans, polycyclic aromatic hydrocarbons, polychlorinated biphenyls) and heavy metals are quite high in Guiyu as compared with other cities. Concentrations of the aforesaid POPs and heavy metals detected in the Guiyu air are especially high; this is primarily due to incomplete combustion of E-waste scrap. They argue that this contamination of the air leads to the severe pollution of the soil by POPs and heavy metals, which may also affect the surrounding environment (such as rice fields and rivers) through atmospheric movement and deposition.

Under such high concentrations of toxic chemicals, the health conditions of the workers and local residents can be adversely affected through inhalation, dermal exposure, and oral ingestion of contaminated drinking water and food. Researchers from the Department of Medicine in the University of Shantou studied the health of 165 Guiyu children in 2003 and 2004; 135 children were found to be afflicted with lead poisoning (Huo *et al.*, 2007).

7.3 The economic background to the material flows of E-waste

7.3.1 The global picture of the material flows of E-waste

As stated in the Introduction, the global picture of E-waste material flows can be split into two closed regions. The first region includes the western half of the US, Japan, Korea, China, Hong Kong, and Southeast Asian countries. The end-of-life EEE generated in the US, Japan, Korea, Hong Kong, Malaysia, and Singapore are exported as second-hand items to other countries, especially Southeast Asian countries and China. But E-waste scrap, especially PCBs, is imported mainly by China.

We also find a similar structure of resource circulation in the second region. In this region, while the EU and the eastern half of the US export second-hand items to Africa, the Middle East, and South Asia, E-waste scrap is imported mainly by India and Pakistan. Why are China, India, and Pakistan such intensive centers of E-waste recycling?

To examine the key economic drivers that make the resource circulation in the world what it is today, we have to look at demands for second-hand items and for E-waste scrap separately.

7.3.2 The demand side for second-hand goods

We investigated the second-hand markets in some developing countries in the course of seven field studies from 2005 to 2010.[3] The size of second-hand markets varies from country to country. While the markets in Thailand and Nepal are very thin, those in India, Vietnam, and the Philippines are huge. Figure 7.1 shows GDP per capita in some Asian countries. The figure tells us that there is a demand for second-hand items only in middle-income developing countries. We

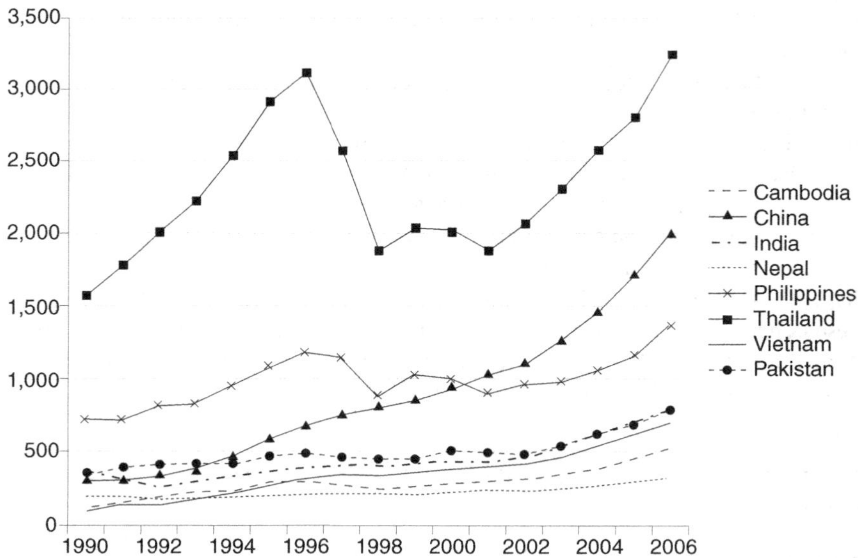

Figure 7.1 GDP per capita in some Asian countries (source: Development Data Group, The World Bank, 2008. 2008 World Development Indicators Online. Washington, DC: The World Bank. Available at: http://go.worldbank.org/U0FSM7AQ40).

can hypothesize that there is a demand for second-hand items in countries in which the GDP per capita ranges from $500 to $2,000.

Although we did not visit other regions like Africa and South America, we can predict the second-hand markets in those regions. The hypothesis stated above implies that Sub-Saharan African countries will become huge markets for second-hand items in the near future. In 2000 the number of countries potentially demanding second-hand items was nine, but it totaled 15 in 2006.[4] In the Sub-Saharan African region, the GDP per capita remained low, on average around the lower limit of $500, until 2002, after which the average GDP per capita has been dramatically rising (from $495 in 2002 to $920 in 2006).

Similarly, according to that hypothesis, there is not going to be a large market in South America for second-hand items. Because in 2006 there are only three countries—Bolivia, Guyana, and Paraguay—where the per capita GDP is within the range of $500 to $2,000,[5] we can safely surmise that the second-hand market in this region will not be large in the near future.

7.3.3 The demand side for scrap

Scrap is mainly imported by developing countries, and for several reasons. First, developed countries specialize in the production of goods with high added value and they use virgin resources to maintain the quality of those products. Second, the price of scrap is cheaper than that of virgin resources. Third, recycling or resource recovery from scrap is quite labor intensive, because recycling requires scrap to be sorted out by each different type of material, and so it cannot be completely done by any machines.

The demand for scrap in a developing country increases with that country's GDP. The top six developing countries in 2006 were China, Brazil, India, Korea, Mexico, and Turkey, as shown in Table 7.2. In addition, since the industry of recycling is labor intensive compared with other industries, countries with abundant cheap labor enjoy an advantage in the industry. Table 7.2 also includes the ratio of the population living on less than $2 per day. The table suggests that China, India, and Pakistan will be potentially the centers of E-waste recycling in the world.

7.3.4 A comparison between reuse value and recycling value

Finally, let us look more closely at the material flows of E-waste. To do this, we need to split the value of second-hand items exported to developing countries into their reuse value and their scrap value. Table 7.1 indicates that the maximum benefit accrued from recycling a personal desktop computer is $54.64 per computer (calculated by adding the sums given in the last column in the table). This value is the maximum scrap value of a second-hand PC, since it is calculated under the assumption that the recycling rate is 100 percent and the recycling cost is zero. The calculation also assumes that other materials cannot be economically recovered.

Table 7.2 GDP of developing countries in 2006 and population living on less than $2/day

Country	*GDP in 2006 (million current US$)*	*Population living on less than $2/day*
China	2,644,681 (1)	34.9% in 2004
Brazil	1,067,472 (2)	21.2% in 2004
India	911,813 (3)	80.4% in 2004
Korea	888,024 (4)	2% in 1998
Mexico	839,182 (5)	11.6% in 2004
Turkey	402,710 (6)	18.7% in 2003
...	...	...
Pakistan	126,836 (18)	73.6% in 2002

Sources: Development Data Group, The World Bank, 2008. 2008 World Development Indicators Online. Washington, DC: The World Bank. Available at: http://go.worldbank.org/U0FSM7AQ40. Development Data Group, The World Bank, 2007. 2007 World Development Indicators Online. Washington, DC: The World Bank. Available at: http://go.worldbank.org/3JU2HA60D0.

Note

The numbers in parentheses denote the order of scale of the GDP. "Population living on less than $2/day" is the percentage of the population living on less than $2.15 a day at 1993 international prices (equivalent to $2 in 1985 prices, adjusted for purchasing power parity).

Table 7.3 A price list of second-hand PCs in Sharja

Producer	*Specifications*	*Price (current US$)*
Compaq	Dsk, Pentium4, 2.0GHz, 512MB, 40GB	$81 (300 AED)
Compaq	Dsk, Pentium4, 3.0GHz, 1GB, 250GB	$270 (1,000 AED)
HP	Dsk, Pentium4, 3.0GHz, [xw 4300 workstation]	$283.50 (1,050 AED)
HP	Dsk, Core2duo, 2.3GHz, [xw 4600 workstation]	$486 (1,800 AED)
HP	Dsk, Pentium4, 3.2GHz, 512MB, 40GB	$70.20 (260 AED)
HP	Dsk, Pentium4, 2.3GHz, 1GB, 160GB	$128.30 (475 AED)
HP	PC monitor, LCD, 15th inch	$78.30 (290 AED)
HP	PC monitor, LCD, 15th inch	$81 (300 AED)
Dell	PC monitor, LCD, 17th inch	$148.50 (550 AED)

Source: The field survey in Sharja by Shinkuma with Keisaku Higashida from November 1 to 5, 2009.

Note

1 AED = $0.27 (November 1, 2009).

On the other hand, Tables 7.3 and 7.4 show lists of the selling prices for second-hand PCs and PC monitors in Sharja in the UAE and Manila in the Philippines, respectively. The average price of a desktop PC in Sharja and Manila is $146.90 ($219.80 in Sharja and $98.30 in Manila). Therefore, the reuse value of a second-hand PC is greater by $92.26: the average total value of a second-hand PC ($146.90) minus the scrap value ($54.64). These simple calculations show that the profit from recycling is not as large as that from selling items for reuse, and this provides developed countries with the motivation to export the end-of-life items as second-hand items rather than as E-waste scrap.

Table 7.4 A price list of second-hand PCs and monitors in Manila

Producer	*Specifications*	*Price (current US$)*
Samsung	Dsk, Pentium4, 2.4GHz, 512MB, 40GB	$104.30 (4,700 PHP)
Samsung	Dsk, Pentium4, 2.8GHz, 512MB, 40GB	$111 (5,000 PHP)
Samsung	Dsk, Pentium4, 2.4GHz, 512MB, 40GB	$138.70 (6,249 PHP)
Samsung	Dsk, Pentium4, 1.7GHz, 256MB, 40GB	$64.40 (2,900 PHP)
Samsung	Dsk, Pentium4, 2GHz, 512MB, 40GB	$66.60 (3,000 PHP)
Samsung	Dsk, Pentium4, 2.6GHz, 512MB, 40GB	$88.80 (4,000 PHP)
LG	Dsk, Pentium4, 2.4GHz, 512MB, 40GB	$77.70 (3,500 PHP)
LG	Dsk, Pentium4, 2.8GHz, 512MB, 80GB	$111 (5,000 PHP)
LG	Dsk, Pentium4, 3GHz, 512MB, 80GB	$122.10 (5,500 PHP)
Samsung	PC monitor, CRT, 15th inch	$48.80 (2,200 PHP)
Samsung	PC monitor, LCD, 15th inch	$93.20 (4,200 PHP)
Samsung	PC monitor, CRT, 17th inch	$33.30 (1,500 PHP)
Logix	PC monitor, CRT, 15th inch	$28.70 (1,300 PHP)
Dell	PC monitor, CRT, 17th inch	$33.30 (1,500 PHP)
Dell	PC monitor, CRT, 15th inch	$24.40 (1,100 PHP)
Dell	PC monitor, LCD, 15th inch	$77.70 (3,500 PHP)
Dell	PC monitor, LCD, 17th inch	$99.90 (4,500 PHP)

Source: The field survey in Manila by Shinkuma with Keisaku Higashida from February 13 to 20, 2010.

Note
1 PHP = $0.0222 (February 18, 2010).

In actual fact, however, it is not only second-hand items that are exported from developed countries but also E-waste scrap. It is the recycling systems in those developed countries that make the exporting of scrap profitable. Note that recyclers in developed countries are subsidized for collecting end-of-life goods and roughly sorting the scrap into each material by machines. The subsidies are usually funded by the fees paid by producers or consumers.

In developing countries that have not yet introduced any recycling systems, however, the domestic recyclers or the importers of scrap must pay the price corresponding to the scrap value of end-of-life goods to the original owner. The relatively low profitability of recycling means that the scrap will be recycled within or near the developing countries where the second-hand items were originally used.

There are several huge second-hand markets around the big recycling centers located in China, India, and Pakistan. This fact is not a geographical accident. Consumers near each of those big recycling centers can sell second-hand goods as scrap after using those goods. This implies that the net price that those consumers will pay for the second-hand goods is lower by the scrap price that they can receive from recyclers.

Around China there are several huge second-hand markets in Southeast Asia, such as Vietnam and the Philippines. China itself is also a large second-hand market. Hong Kong appears to be another gateway for second-hand goods to enter China. Similarly, India and Pakistan are themselves huge second-hand markets, with several second-hand markets in places like the Middle East and western Africa which are not so far from the Indian subcontinent. Just as China enjoys the benefit of having a well-developed gateway in Hong Kong, so India and Pakistan can enjoy the benefit of Sharja in the UAE, for through it pass many goods from the EU and the eastern half of the US.

7.4 International trade data supporting the material flows of E-waste

In this section we examine by means of international trade data the global picture of the material flows discussed in the preceding section. Figure 7.2 below shows the time trend of trading in two metal scraps: copper and ferrous metal. It shows that the world trade in these two items has increased since 2000, and this seems to show that the trading of E-waste scrap has also increased. Is this true, though?

Unfortunately, there does not exist any HS code assigned to E-waste scrap. As a second-best strategy to answer the above question, we choose the alternative of looking at "other ferrous scraps" (720449 in the HS classification). The reasons we focus on the HS six-digit code of 720449 are as follows. First of all, this type of ferrous scrap usually includes other metals such as copper, aluminum, and lead. Second, at least in Japan E-waste scrap has been exported as this type of scrap.[6] Third, as shown in Table 7.1, iron is the major metallic component of E-waste scrap by weight, and it is followed by aluminum, copper, and lead.

Tables 7.5, 7.6, 7.7, and 7.8 show the amount of "other ferrous" exported to China, Hong Kong, India, and Pakistan, respectively. They are completely consistent with our sketch of the material flows of E-waste scrap. Table 7.5 shows that China is the recycling center of the first region. From Tables 7.5 and 7.6 we can conclude that Hong Kong is the second entrance to China. In the first region,

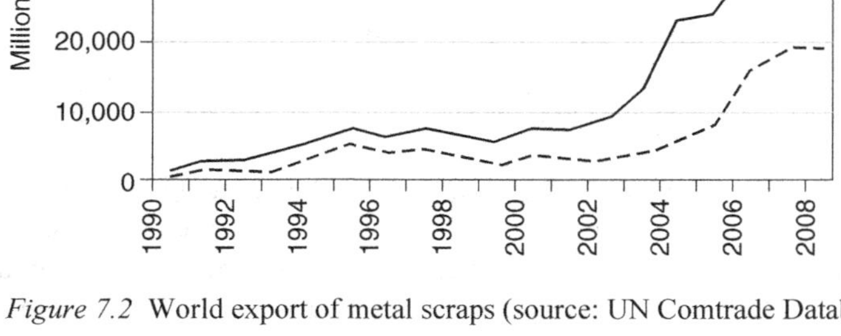

Figure 7.2 World export of metal scraps (source: UN Comtrade Database).

Table 7.5 The export of "other ferrous" (HS: 720449) to China in 2007

From	*Quantity (tons)*
Japan	2,160,709
United States	1,167,961
Hong Kong	532,501
EU15	183,863
Canada	91,543
Thailand	81,404
Russia	51,163
Korea	21,927
Singapore	21,635
Australia	17,525
Philippines	16,817
...	...
India	48

Source: World Trade Atlas.

Table 7.6 The export of "other ferrous" (HS: 720449) to Hong Kong in 2007

From	*Quantity (tons)*
United States	54,936
Canada	49,487
EU15	23,621
Japan	21,430
Taiwan	5,366
Singapore	3,548
Philippines	2,490
Korea	2,223
Mexico	1,383

Source: World Trade Atlas.

Japan and the western half of the US are the main exporters. It should be noted that the EU also exports "other ferrous" scrap to China. This fact implies that the scrap demand in China is larger than that in India (see also Table 7.7). Another thing to note is that the Philippines exports the scrap to China, which may support our prediction that second-hand items are reused in Southeast Asian countries such as the Philippines and then the E-waste scrap is recycled in China.

According to Tables 7.7 and 7.8, the eastern half of the US and the EU are the major exporters to India and Pakistan, which are the main recycling centers of the second region. It should be noted that South Africa exports "other ferrous" scraps to those countries, which may imply that second-hand items are reused in east African countries, including South Africa, and the scrap is recycled in India

Table 7.7 The export of "other ferrous" (HS: 720449) to India in 2007

From	*Quantity (tons)*
United States	534,912
EU15	207,322
Canada	201,002
Singapore	45,249
Japan	6,806
Philippines	5,526
Australia	5,058
Korea	4,560
South Africa	3,908
Lithuania	3,651
Thailand	2,946
...	...
China	0

Source: World Trade Atlas.

Table 7.8 The export of "other ferrous" (HS: 720449) to Pakistan in 2007

From	*Quantity (tons)*
EU15	233,030
United States	142,857
South Africa	25,278
Canada	15,363
Singapore	13,520
Japan	2,630
Romania	2,411
Russia	2,204

Source: World Trade Atlas.

and Pakistan. In addition, Tables 7.5 and 7.7 show that China, India, and Pakistan are independent recycling centers in that little trading of scrap seems to take place among them.

Tables 7.9 and 7.10 respectively show the amount of second-hand cathode-ray tube (CRT) television sets and second-hand PC monitors exported from Japan in 2005.[7] According to these tables, huge amounts of second-hand items are exported from Japan to Southeast Asian countries and China. In addition, bear in mind that Hong Kong serves as a distributor between Japan and other Asian countries. All the above international trade data seems to support our global picture of the material flows of E-waste.

The international trade in hazardous waste such as E-waste has been regulated by the Basel Convention. In Chapter 9 we shall examine the effectiveness of the Basel Convention.

Table 7.9 The amount of second-hand CRT television sets exported from Japan in 2005

Hong Kong	2,835,668
Philippines	658,293
Cambodia	281,124
Malaysia	167,723
Myanmar	166,010
China	92,779
Afghanistan	20,156
Others	45,259

Source: The original data from Japan Environmental Sanitation Center *et al.* (2007). This table is a reprint of Table 2 in Shinkuma and Huong (2009).

Table 7.10 The amount of second-hand personal computer monitors exported from Japan in 2005

Hong Kong	1,354,963
Vietnam	197,815
Thailand	65,323
Arab countries	27,264
Malaysia	14,440
China	13,011
Cambodia	8,598
Indonesia	6,643
Others	18,015

Source: The original data from Japan Environmental Sanitation Center *et al.* (2007). This table is a reprint of Table 3 in Shinkuma and Huong (2009).

8 The effects of introducing EPR policies in developed countries on international trade, the environment, and social welfare[1]

8.1 Introduction

As we have seen already in this book, EPR-based policies have been adopted with regard to end-of-life goods, especially in developed countries. They are classified into two types: a product take-back requirement and deposit/refund systems, where the latter includes a deposit/refund system (in a narrow sense), a UCTS, and an ADF policy (see Section 3.4). One of the common features in these policies is that, in order to prevent illegal disposal, consumers are not required to pay a disposal fee or the costs of recycling end-of-life goods when they dispose of them.

In Chapter 6 we pointed out one of the defects of EPR policies: they are not socially optimal when they are adopted for durable goods rather than nondurable goods. We look at another defect of EPR policies in this chapter. The scope of EPR policies in a developed country is narrow, in the sense that it does not duly take into account the effect on international trade, on the environment outside the country, or on social welfare—especially the social welfare of developing countries.

It is important to keep in mind that recycling in a developed country is not closed. This is true especially in the case of end-of-life durable goods, which can be exported from developed countries to developing countries as second-hand goods.[2] They are also exported from developed countries as scrap to developing countries such as China and India, and it is the recovery of metals from such scrap that seriously pollutes the environment.

The level of damage caused to the environment and to social welfare through international trade depends on the policies that are adopted in the exporting and importing countries. In this chapter we set out to investigate the effects that introducing an EPR policy in a developed country will have, or can have, on international trade, the environment, and social welfare. What we shall find is that adopting an EPR policy in a developed country cannot be *globally* optimal. In addition, adoption of an EPR policy in a developed country will necessarily decrease social welfare in a developing country and increase the amount and severity of environmental damage in the latter.

Studies on trade and environment draw as their main conclusion that trade can lower welfare in the absence of an adequate environmental policy but trade always raises welfare with an efficient environmental policy (Copeland, 1994; Krutilla, 1991).

There are few theoretical studies on waste disposal or recycling and trade with the exception of those done by Copeland (1991) and Cassing and Kuhn (2003). In a small open economy model, Copeland (1991) shows that the first-best policy is to allow free trade in waste disposal and to correct the externality with domestic environmental policies, including a tax on waste disposal activity. This implies that if the domestic environmental policy is set optimally, no tax on imported waste is necessary. He also shows that, if the first-best policy is for some reason not feasible, trade restriction in the area of waste disposal services can improve welfare as a supplement to the domestic environmental policy. Our aim here, however, is to examine the effect of a change in policy in an exporting country on the environment and welfare of an importing country.

Cassing and Kuhn (2003) deal with international trade in hazardous waste when there is an international oligopoly market for waste, and the governments of both waste-importing and waste-exporting countries act strategically by implementing national environmental policies. They analyze how national environmental policies are distorted when seen from the point of view of the global optimum. While "waste" in their model is defined as a by-product of production, in our model it is defined in terms of end-of-life durable goods.

Although many researchers in environmental science are paying careful attention to the E-waste problem, and it has been widely recognized as one of the most serious of environmental problems, there are few theoretical models that deal with this problem directly.

Several studies on recycling and trade examine other topics. Grace *et al.* (1978) develop a model of trade in secondary materials and refer to some policy aspects. Van Beukering and Bouman (2001) examine patterns of trade in recyclable materials using the Hechsher–Ohlin model. Huhtala and Samakovlis (2002) examine standards applied to recycling activities. Higashida and Jinji (2006) examine the strategic aspects of recycled content standards.

The outline of this chapter is as follows. In Section 8.2 we derive a social optimum as a benchmark. In Section 8.3 we compare the competitive equilibrium under an EPR policy with the social optimum. Section 8.4 summarizes this chapter.

8.2 Social optimum

We consider an open economy consisting of a developed country (country N) and a developing country (country S). Both countries consume durable goods. They trade second-hand goods and scrap.

In order to focus on international trade in second-hand goods and scrap between a developed country and a developing country, we assume that the unit production cost of a durable good is constant and it is the same value across the two countries. It is denoted by P (constant).

In this section, we derive the social optimum. We assume that there does not exist any demand for second-hand goods in country N, but there is in country S. A representative consumer in country N uses such goods for only one period. Let x and $u_N(x)$ stand for consumption of new goods and utility from the consumption. It is assumed that $u'_N(x) > 0$ and $u''_N(x) < 0$. In the next period they become end-of-life durable goods in country N. However, end-of-life durable goods can be exported to country S as second-hand goods. We assume that it costs to export second-hand goods. The exporting cost is represented by $d(x_u)$, where x_u denotes the volume of exported second-hand goods. We assume that $d'(x_u) > 0$ and $d''(x_u) > 0$.

The residual of end-of-life durable goods $x - x_u$ is recycled domestically (x_s) or exported to country S as scrap ($x - x_u - x_s$). Hence, we assume that country N (S) exports (imports) both second-hand goods and scrap.[3] The recycling cost in country N is represented by $C_N(x_s)$ and it is assumed that $C'_N(x_s) > 0$ and $C''_N(x_s) > 0$. A unit constant cost denoted by q for scrapping and exporting is assumed. The constant value of recovered resource is denoted by p_m.

Next, we consider country S, which imports second-hand goods and scrap. We assume that, in country S, a representative consumer gains some positive utility not only from new goods denoted by y but also from second-hand goods. The utility from the consumption of new goods is represented by $u_S(y)$. Note that the consumption of second-hand goods is represented by $y + y_u$ in the steady state of our economy, where the imported second-hand goods are denoted by y_u. The utility gained from the consumption of second-hand goods is represented by $v(y + y_u)$. We assume that u_s and v are concave functions. A second-hand good becomes, in the next period, an end-of-life durable good in country S. Such goods are recycled domestically by a representative recycler.

The recycling cost in country S is represented by $C_S(y_s)$, where y_s denotes the end-of-life durable goods to be recycled. Since country S imports scrap from country N, scrap imported is included in y_s. It is assumed that $C'_S(\cdot) > 0$ and $C''_S(\cdot) > 0$.

For simplicity, we assume that recycling in country N does not pollute the environment but that in country S it does. We denote by E the constant marginal environmental cost associated with recycling.

Taking this environmental cost into account, we can represent the total sum of national welfares in the world as

$$SW = u_N(x) + u_S(y) - P(x + y) + v(y + y_u) - d(x_u)$$

$$+ p_m(x_s + y_s) - C_N(x_s) - C_S(y_s) - q(x - x_u - x_s) - Ey_s.$$

Note that the social optimum can be obtained so that social welfare is maximized subject to the following constraints:

$$x_u = y_u \tag{8.1}$$

$$x - x_u - x_s = y_s - y - y_u. \tag{8.2}$$

These equations represent the material balance conditions for second-hand goods and scrap respectively. The FOCs for maximization of *SW* can be obtained as follows:

$$u'_N(x) - P - (q - p_s) = 0 \quad (8.3)$$

$$p_u + (q - p_s) - d'(x_u) = 0 \quad (8.4)$$

$$p_m + (q - p_s) - C'_N(x_s) = 0 \quad (8.5)$$

$$u'_S(y) - P + v'(y + y_u) + p_s = 0 \quad (8.6)$$

$$v'(y + y_u) + p_s - p_u = 0 \quad (8.7)$$

$$p_m - C'_S(y_s) - p_s - E = 0. \quad (8.8)$$

Note that p_u and p_s represent the shadow price of second-hand goods and scrap respectively or Lagrange multipliers associated with constraints (8.1) and (8.2).

We introduce the new variables $\hat{p}$ and $\tilde{p}$ defined by

$$\hat{p} = P - p_s$$

$$\tilde{p} = p_u - p_s,$$

respectively. Using these variables, we can derive the socially optimal value x^*, x^*_u, x^*_s, y^*, y^*_u, and y^*_s as follows:

$$x^* = u'^{-1}_N(\hat{p} + q) \quad (8.3)'$$

$$x^*_u = d'^{-1}(\tilde{p} + q). \quad (8.4)'$$

$$x^*_s = C'^{-1}_N(\hat{p} - P + p_m + q) \quad (8.5)'$$

$$y^* = u'^{-1}_S(\hat{p} - \tilde{p}) \quad (8.6)'$$

$$y^* + y^*_u = v'^{-1}(\tilde{p}) \quad (8.7)'$$

$$y^*_s = C'^{-1}_S(\hat{p} - P + p_m - E). \quad (8.8)'$$

Substituting these equations into material balance conditions (8.1) and (8.2), we obtain

$$d'^{-1}(\tilde{p} + q) = v'^{-1}(\tilde{p}) - u'^{-1}_S(\hat{p} - \tilde{p}) \quad (8.9)$$

$$u'^{-1}_N(\hat{p} + q) - d'^{-1}(\tilde{p} + q) - C'^{-1}_N(\hat{p} - P + p_m + q) = C'^{-1}_S(\hat{p} - P + p_m - E) - v'^{-1}(\tilde{p}). \quad (8.10)$$

8.3 The effect of an EPR policy in country N

8.3.1 The international trade equilibrium if country N introduces an EPR policy

Now suppose that country N is going to introduce an EPR policy. In the following subsections we investigate the effect of an EPR policy in country N on the international trade equilibrium, the environment, and welfare. In this subsection we derive the international trade equilibrium when country N introduces an EPR policy. We assume that both governments do not behave strategically and consider prices as given. Equivalently, it is assumed that both governments do not consider the effects of their policies on terms of trade. We have considered that EPR policies are best represented by an ADF policy in Chapter 6, so we use an EPR and an ADF policy interchangeably.

Under an EPR (or an ADF) policy, a consumer must pay the recycling fee for an end-of-life durable good in advance when it is first purchased. We denote by r the advance fee. Note that the advance fee is determined in the competitive market.

The consumer surplus of a representative consumer in country N can be written as

$$\underset{x,\,x_u}{Max}\ \ CS_N = u_N(x) - (P + r)x + p_u x_u - d(x_u).$$

The representative consumer maximizes the consumer surplus with respect to x and x_u. Note that the second-hand-good price is denoted by p_u. The FOCs of the problem stated above can be given as

$$u'_N(x) - P - r = 0 \tag{8.11}$$

$$p_u - d'(x_u) = 0. \tag{8.12}$$

A representative recycler in country N receives end-of-life durable goods denoted by x_p from the consumer in exchange for the competitive recycling fee denoted by r. A part of end-of-life durable goods denoted by x_s is recycled domestically but the residual is exported as scrap. Then we can state the profit maximization problem of the recycler as follows:

$$\Pi_N = p_m x_s - C_N(x_s) + r x_p + (p_s - q)(x_p - x_s).$$

Note that the scrap price is denoted by p_s. The FOCs can be given as

$$p_m - C'_N(x_s) - (p_s - q) = 0 \tag{8.13}$$

$$r + p_s - q = 0. \tag{8.14}$$

Competitive market equilibrium for x, x_u, and x_s can be obtained from (8.11) to (8.14) as follows:

$$x^* = u_N'^{-1}(\hat{p} + q) \tag{8.11)'$$

$$x_u^* = d'^{-1}(\tilde{p} - \hat{p} + P) \tag{8.12)'$$

$$x_s^* = C_N'^{-1}(\hat{p} - P + p_m + q). \tag{8.13)'$$

Note that $x_p^* = x^* - x_u^*$ in the competitive equilibrium. Comparing (8.3)' to (8.5)' with (8.11)' to (8.13)', we can state that an EPR policy cannot produce the social optimum. This result is summed up in the following proposition:

Proposition 8.1. *An EPR policy is not the first-best policy for a country exporting second-hand goods and scrap.*

Since recycling in country S generates an environmental pollution, country S needs to implement some policies. In order to focus on the effect of the adoption of an EPR policy in country N on international trade equilibrium, as a benchmark we consider the case where country S adopts the first-best policy. We consider a tax levied on consumption of new goods denoted by t_n and tariff on imported second-hand goods and imported scrap denoted by t_u and t_s respectively. The consumer surplus of the representative consumer in country S can be written as

$$CS_S = u_S(y) - (P + t_n)y + v(y + y_u) - (p_u + t_u)y_u + (p_s + t_s)(y + y_u).$$

Note that the domestic prices of second-hand goods and scrap are represented by $p_u + t_u$ and $p_s + t_s$ respectively. The representative consumer maximizes the consumer surplus with respect to y and y_u. The FOCs for the maximization problem can be written as

$$u'_S(y) - P - t_n + v'(y - y_u) + p_s + t_s = 0 \tag{8.15}$$

$$v'(y + y_u) - p_u - t_u + p_s + t_s = 0. \tag{8.16}$$

The profit of the recycler in country S is represented as

$$\Pi_S = (p_m - p_s - t_s)y_s - C_S(y_s).$$

The FOC for the profit maximization problem can be obtained:

$$p_m - p_s - t_s - C'_S(y_s) = 0. \tag{8.17}$$

By a simple arrangement of (8.15) to (8.17) we can obtain market equilibrium value of y, y_u, and y_s as follows:

$$y^* = u_S'^{-1}(\hat{p} - \tilde{p} + t_n - t_u) \tag{8.15)'$$

$$y^* + y^*_u = v'^{-1} (\tilde{p} + t_u - t_s) \tag{8.16)'}$$

$$y^*_s = C'^{-1}_S (\hat{p} - P + p_m - t_s). \tag{8.17)'}$$

Comparing these equations with equations (8.6)' to (8.8)' respectively, we can see that the optimal policy is characterized by

$$t_n = t_u = t_s = E. \tag{8.18}$$

In the following subsections we assume that the government of country S adopts the optimal policy described in (8.18).

Substituting (8.11)', (8.12)', (8.13)', (8.15)', (8.16)', and (8.17)' into (8.1) and (8.2), we obtain

$$d'^{-1} (\tilde{p} - \hat{p} + P) = v'^{-1} (\tilde{p}) - u'^{-1}_S (\hat{p} - \tilde{p})\, u'^{-1}_N \tag{8.19}$$

$$(\hat{p} + q) - d'^{-1} (\tilde{p} - \hat{p} + P) - C'^{-1}_N (\hat{p} - P + p_m + q)$$
$$= C'^{-1}_S (\hat{p} - P + p_m - E) - v'^{-1} (\tilde{p}). \tag{8.20}$$

Note that equations (8.1) and (8.2) also represent market clearance conditions for second-hand goods and scrap respectively.

8.3.2 The effect of adoption of an EPR policy by country N on the international trade equilibrium and the environment

In this section we examine the effect of adoption of an EPR policy by country N on international trade by comparing (8.9) and (8.10) with (8.19) and (8.20). Note that by substituting (8.9) (or (8.19)) into (8.10) (or (8.20)) the following equation can be obtained:

$$u'^{-1}_N (\hat{p} + q) + u'^{-1}_S (\hat{p} - \tilde{p}) - C'^{-1}_N (\hat{p} - P + p_m + q)$$
$$- C'^{-1}_S (\hat{p} - P + p_m - E) = 0. \tag{8.21}$$

This equation states that there exists a one-to-one correspondence between $\hat{p}$ and $\tilde{p}$ and that the correspondence does not depend on whether an EPR policy is implemented in country N or not. By applying the implicit function theorem for (8.21), we obtain

$$\frac{\partial \tilde{p}}{\partial \hat{p}} = 1 + \frac{u''_S}{u''_N} - \frac{u''_S}{C''_S} - \frac{u''_S}{C''_N} > 1. \tag{8.22}$$

From equation (8.22), the following lemma can be obtained.

Lemma 8.1 *If and only if $\hat{p}_{EPR} < \hat{p}_{OP}$, then $\tilde{p}_{EPR} < \tilde{p}_{OP}$. In addition, if and only if $\hat{p}_{EPR} > \hat{p}_{OP}$, $\hat{p}_{EPR} - \tilde{p}_{EPR} < \hat{p}_{OP} - \tilde{p}_{OP}$. The subscripts EPR and OP represent an EPR policy and the social optimum respectively.*

Proof. The first statement is clear from (8.22). The second statement can be proved as follows. Suppose that $\hat{p}_{EPR} > \hat{p}_{OP}$. From equation (8.22), we can easily see that $\hat{p}_{EPR} - \hat{p}_{OP} < \tilde{p}_{EPR} - \tilde{p}_{OP}$. Alternatively, suppose that $\hat{p}_{EPR} < \hat{p}_{OP}$. Then from equation (8.22), $\hat{p}_{OP} - \hat{p}_{EPR} < \tilde{p}_{OP} - \tilde{p}_{EPR}$.

By comparing (8.9) and (8.10) with (8.19) and (8.20), we obtain the following proposition.

Proposition 8.2. *(i) Suppose that $q > p_s$. Then if country N adopts an EPR policy, both $\hat{p}$ and $\tilde{p}$ increase. In other words, the inequalities $\tilde{p}_{OP} < \tilde{p}_{EPR}$ and $\hat{p}_{OP} < \hat{p}_{EPR}$ hold. (ii) Suppose that $q < p_s$. Then if country N adopts an EPR policy, both $\hat{p}$ and $\tilde{p}$ decrease.*

Proof. We prove for case (i). (For the proof of the opposite case see Appendix 8.) Suppose that $\tilde{p}_{OP} + q < \tilde{p}_{EPR} - \hat{p}_{EPR} + P$. Then the l.h.s. of (8.19) is larger than that of (8.9), since $d'^{-1\prime} = \frac{1}{d''} > 0$. Therefore, the following inequality holds:

$$v'^{-1}(\tilde{p}_{OP}) - u'^{-1}_{S}(\hat{p}_{OP} - \tilde{p}_{OP}) < v'^{-1}(\tilde{p}_{EPR}) - u'^{-1}_{S}(\hat{p}_{EPR} - \tilde{p}_{EPR}). \tag{8.23}$$

Suppose that $\tilde{p}_{OP} < \tilde{p}_{EPR}$. Then from Lemma 8.1, $\hat{p}_{EPR} - \tilde{p}_{EPR} < \hat{p}_{OP} - \tilde{p}_{OP}$. Note that $u'^{-1\prime}_{S} = \frac{1}{u''_{S}} < 0$ and $v'^{-1\prime} = \frac{1}{v''} < 0$. We can easily see that inequality (8.23) does not hold, since $v'^{-1}(\tilde{p}_{OP}) > v'^{-1}(\tilde{p}_{EPR})$ and $u'^{-1}_{S}(\hat{p}_{OP} - \tilde{p}_{OP}) < u'^{-1}_{S}(\hat{p}_{EPR} - \tilde{p}_{EPR})$. This is a contradiction. Therefore, $\tilde{p}_{OP} > \tilde{p}_{EPR}$ holds, if

$$\tilde{p}_{OP} + q < \tilde{p}_{EPR} - \hat{p}_{EPR} + P = \tilde{p}_{EPR} + p_s.$$

However, this is also a contradiction, since by our supposition $q > p_s$. Therefore, we can conclude that

$$\tilde{p}_{OP} + q > \tilde{p}_{EPR} - \hat{p}_{EPR} + P. \tag{8.24}$$

If inequality (8.24) holds, the l.h.s. of (8.9) is larger than that of (8.19). Then the following inequality holds:

$$v'^{-1}(\tilde{p}_{OP}) - u'^{-1}_{s}(\hat{p}_{OP} - \tilde{p}_{OP}) > v'^{-1}(\tilde{p}_{EPR}) - u'^{-1}_{s}(\hat{p}_{EPR} - \tilde{p}_{EPR}). \tag{8.25}$$

Suppose that $\tilde{p}_{OP} > \tilde{p}_{EPR}$. Similarly, from Lemma 8.1, $\hat{p}_{EPR} - \tilde{p}_{EPR} > \hat{p}_{OP} - \tilde{p}_{OP}$. This supposition, however, violates inequality (8.25). This is a contradiction. Therefore, we obtain the desired results: $\tilde{p}_{OP} < \tilde{p}_{EPR}$ and $\hat{p}_{OP} < \hat{p}_{EPR}$.

From the theoretical point of view, both case (i) and case (ii) are possible. According to (8.14), however, $r = q - p_s$. In actuality, the price of end-of-life durable goods (r) is negative in developed countries. This is the reason why some of them have been disposed of illegally and why an ADF system (or an EPR policy) has been introduced to prevent illegal disposal. Therefore, we focus on case (i) and omit the opposite case, although we can easily see how the results are modified in the opposite case.

Lemma 8.1 and Proposition 8.2 (i) state that second-hand-good price p_u increases but scrap price p_s decreases if country N introduces an EPR policy. To see this, suppose that country N applies a traditional disposal fee (DF) system instead of an ADF system. In the former system, the consumer is required to pay a recycling fee when he/she disposes of end-of-life durable goods. In addition, it can be easily shown that a DF system attains the social optimum when illegal disposal is not a possibility.

When an ADF policy is replaced by a DF policy, the consumer has a stronger incentive to avoid the payment of recycling fees by exporting end-of-life goods. This is an intuitive reason why a DF policy pushes the second-hand-good price downward or an ADF (or an EPR) policy pushes it upward compared with a DF policy or the social optimum.

In other words, an EPR policy gives the consumer too weak an incentive for exporting second-hand goods to obtain the social optimum. The weak incentive to export end-of-life durable goods as second-hand goods implies that domestic recycling increases and hence scrap to be exported also increases. This effect pushes down scrap price p_s. This is why the second-hand-good price p_u increases but the scrap price p_s decreases if country N adopts an EPR policy.

Applying Proposition 8.2 (i), we can easily see the effect of the adoption of an EPR policy by country N on the competitive equilibrium. These results can be summarized in the following propositions (Propositions 8.3 and 8.4).

Proposition 8.3. *Both the export of second-hand goods and the consumption of new goods decrease in country N but the domestic recycling increases, if country N adopts an EPR policy.*

Proof. We can obtain the following inequalities from Proposition 8.2 (i) and Lemma 8.1 directly comparing (8.3)', (8.4)', and (8.5)' with (8.11)', (8.12)', and (8.13)' respectively.

$$x^*_{(OP)} = u'^{-1}_N (\hat{p}_{OP} + q) > x^*_{(EPR)} = u'^{-1}_N (\hat{p}_{EPR} + q) \tag{8.26}$$

$$x^*_{u(OP)} = d'^{-1} (\tilde{p}_{OP} + q) > x^*_{u(EPR)} = d'^{-1}(\tilde{p}_{EPR} - \hat{p}_{EPR} + P) \tag{8.27}$$

$$x^*_{s(OP)} = C'^{-1}_N (\hat{p}_{OP} - P + p_m + q) < x^*_{s(EPR)} = C'^{-1}_N (\hat{p}_{EPR} - P + p_m + q). \tag{8.28}$$

Inequalities (8.26) and (8.28) follow directly from Proposition 8.2 (i). Inequality (8.27) follows from inequality (8.24).

Proposition 8.4. *Suppose that country N adopts an EPR policy. Then the consumption of new goods in country S increases, the consumption of second-hand goods decreases, and recycled scrap increases.*

Proof. Again we look at equations (8.6)', (8.7)', (8.8)', (8.15)', (8.16)', and (8.17)'. Then we can obtain the following inequalities from Proposition 8.2 (i) and Lemma 8.1 directly.

$$y^*_{(OP)} = u'^{-1}_S\,(\hat{p}_{OP} - \tilde{p}_{OP}) < y^*_{(EPR)} = u'^{-1}_S\,(\hat{p}_{EPR} - \tilde{p}_{EPR}) \tag{8.29}$$

$$y^*_{(OP)} + y^*_{u(OP)} = v'^{-1}(\tilde{p}_{OP}) > y^*_{(EPR)} + y^*_{u(EPR)} = v'^{-1}(\tilde{p}_{EPR}) \tag{8.30}$$

$$y^*_{s(OP)} = C'^{-1}_S\,(\hat{p}_{OP} - P + p_m - E) < y^*_{s(EPR)} = C'^{-1}_S\,(\hat{p}_{EPR} - P + p_m - E). \tag{8.31}$$

Inequality (8.29) follows from Proposition 8.2 (i) and Lemma 8.1. The second and last inequalities follow from Proposition 8.2 (i).

This proposition states that damage to the total environment necessarily increases by the adoption of EPR by country N. This result follows from the effect that scrap price p_s decreases.

8.3.3 The effect of an EPR policy in country N on social welfare

To see the effect of EPR adoption in country N on social welfare, we need to obtain some results on comparative statics with respect to the marginal environmental cost E associated with recycling.

We can obtain the next lemma.

Lemma 8.2. *Suppose that the marginal environmental damage E increases. As for the effect on the socially optimal solution, the following results are obtained:*

$$\frac{dx^*_{(op)}}{dE} < 0,\ \frac{dx^*_{u(op)}}{dE} = \frac{dy^*_{u(op)}}{dE} > 0,\ \frac{d(x^*_{(op)} - x^*_{u(op)} - x^*_{s(op)})}{dE}$$
$$= \frac{d(y^*_{s(op)} - y^*_{(op)} - y^*_{u(op)})}{dE} < 0,\ \frac{dy^*_{(op)}}{dE} < 0,\ \textit{and}\ \frac{dy^*_{s(op)}}{dE} < 0.$$

Proof. See Appendix 8.

Now we are ready to investigate the effect that the adoption of EPR by country N will have on the social welfare of each country. We can write the social welfare functions as follows:

$$SW_N = u_N\,(x) - \hat{p}x + \tilde{p}x_u - d(x_u) + (\hat{p} - P + p_m)x_s - C_N\,(x_s) - q(x - x_u - x_s)$$

$$SW_S = u_S\,(y) + v(y + y_u) - \hat{p}y - \tilde{p}y_u + (\hat{p} - P + p_m)y_s - C_S(y_s) - Ey_s.$$

By application of the envelope theorem and equation (8.22), we obtain

$$\frac{\partial SW_N^*}{\partial \hat{p}} = -x_{(op)}^* + x_{u(op)}^* \frac{\partial \tilde{p}}{\partial \hat{p}} + x_{s(op)}^* = -(x_{(op)}^* - x_{u(op)}^* - x_{s(op)}^*) + \left(\frac{\partial \tilde{p}}{\partial \hat{p}} - 1\right) x_{u(op)}^* \tag{8.32}$$

$$\frac{\partial SW_S^*}{\partial \hat{p}} = -y_{(op)}^* + y_{s(op)}^* - y_{u(op)}^* \frac{\partial \tilde{p}}{\partial \hat{p}} = y_{s(op)}^* - y_{(op)}^* - y_{u(op)}^* - \left(\frac{\partial \tilde{p}}{\partial \hat{p}} - 1\right) y_{u(op)}^*. \tag{8.33}$$

Note that we represent by SW_N^* and SW_S^* the maximized social welfare of each country as $\hat{p}$ and $\tilde{p}(\hat{p})$ are given.

Substituting the market clearance conditions represented by (8.1) and (8.2) into these equations, we can easily see that

$$\frac{\partial SW_N^*}{\partial \hat{p}} + \frac{\partial SW_S^*}{\partial \hat{p}} = 0. \tag{8.34}$$

We assume that the marginal environmental cost associated with recycling is sufficiently high. In addition, we assume that $\partial \tilde{p}/\partial \hat{p}$ does not vary largely as E increases.[4] Then, from Lemma 8.2 and (8.34), we can assert that

$$\frac{\partial SW_N^*}{\partial \hat{p}} > 0 \tag{8.35}$$

$$\frac{\partial SW_S^*}{\partial \hat{p}} < 0. \tag{8.36}$$

In regard to the effect of the implementation of an EPR policy in country N on social welfare, we can obtain the following proposition.

Proposition 8.5. *Suppose that country S adopts the first-best policy. In addition, suppose that the marginal environmental cost associated with recycling is sufficiently high. Then the implementation of an EPR policy in country N decreases the social welfare of country S as compared with the social optimal level. This policy change can, however, increase the social welfare of country N.*

Proof. Note that, by Proposition 8.2 (i), $\hat{p}_{OP} < \hat{p}_{EPR}$. Therefore, from (8.36) we can easily see that

$$SW_S^*(\hat{p}_{OP}) > SW_S^*(\hat{p}_{EPR}) = SW_{S(EPR)}.$$

Note that $SW_{S(EPR)}$ represents the social welfare of country S that would be realized in the competitive market if an EPR policy is adopted in country N. In addi-

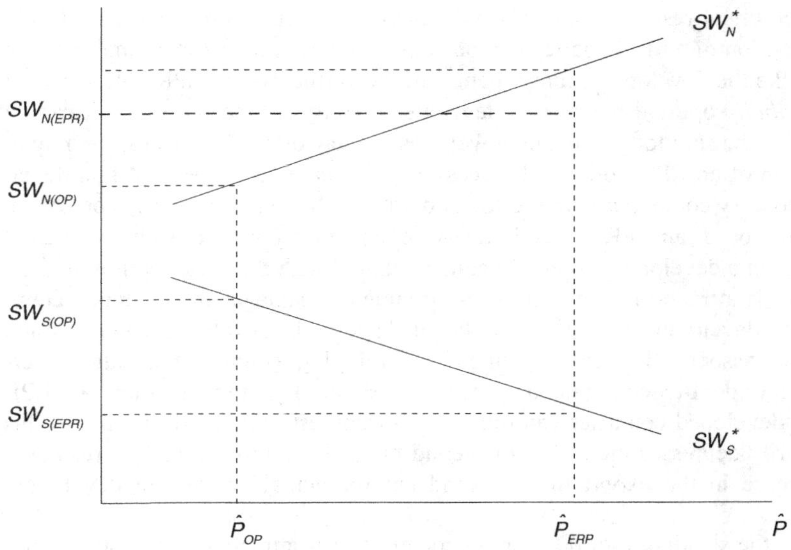

Figure 8.1 The effect of adoption of an EPR policy on social welfare.

tion, note that $SW^*_S(\hat{p}_{EPR}) = SW_{S(EPR)}$, since country S adopts the first-best policy by the supposition.

From (8.35) we can also obtain

$$SW^*_N(\hat{p}_{OP}) < SW^*_N(\hat{p}_{EPR}).$$

However, an EPR policy is not the optimal policy for country N, as stated in Proposition 8.1. Hence, it must be that

$$SW_{N(EPR)} < SW^*_N(\hat{p}_{EPR})$$

Note that $SW_{N(EPR)}$ represents the social welfare of country N that would be realized if an EPR policy is adopted in country N. Therefore, it is ambiguous whether $SW^*_N(\hat{p}_{op}) < SW_{N(EPR)}$ although it is possible.

Figure 8.1 represents the effect of adoption of an EPR policy on social welfare that is stated in Proposition 8.5. This proposition states that the social welfare of country N can increase by the adoption of EPR by country N, if the marginal environmental damage is sufficiently high. However, it should be noted that country N can only increase its social welfare at the expense of the social welfare of country S.

8.4 Summary of this chapter

The theoretical results obtained in this chapter can be summarized as follows. The adoption of an EPR policy can be considered as *nationally* optimal, since it may make the developed country better off. Nevertheless, an EPR policy cannot be a *globally* optimal policy in a developed country, in the sense that it cannot maximize the sum total of national welfares in the world. This means that implementation of an EPR policy will necessarily decrease social welfare in a developing country compared with the social optimum. In addition, it has been shown that adoption of an EPR policy in a developed country increases environmental damages in a developing country, again compared with the social optimum.

There is a trade-off between the environmental damages in developed countries and developing countries (or illegal disposal in developed countries and pollution associated with recycling in developing countries) through international trade of second-hand goods and scrap. By implementing an EPR policy, developed countries can only prevent domestic illegal disposal, but this necessarily depresses the export of second-hand goods too much. This results in an increase in the export of scrap and environmental damage in developing countries.

From the standpoint of developing countries, it is attractive to import second-hand goods rather than scrap. Imported second-hand goods not only give utilities to domestic consumers but also become scrap in the near future. Through the adoption of an EPR policy by developed countries, the environment is polluted severely by recycling much more scrap and domestic consumers lose some of the utility gains that would otherwise be gained from imported second-hand goods.

Appendix

A.8.1 *Proof of Proposition 8.2 (ii)*

Suppose that $\tilde{p}_{OP} + q > \tilde{p}_{EPR} - \hat{p}_{EPR} + P$. Then the l.h.s. of (8.19) is smaller than that of (8.9), since $d'^{-1\prime} = 1/d'' > 0$. Therefore, the following inequality holds:

$$v'^{-1}(\tilde{p}_{OP}) - u'^{-1}_S(\hat{p}_{OP} - \tilde{p}_{OP}) > v'^{-1}(\tilde{p}_{EPR}) - u'^{-1}_S(\hat{p}_{EPR} - \tilde{p}_{EPR}). \tag{A.8.1}$$

Suppose that $\tilde{p}_{OP} > \tilde{p}_{EPR}$. Then from Lemma 8.1, $\hat{p}_{EPR} - \tilde{p}_{EPR} > \hat{p}_{OP} - \tilde{p}_{OP}$. Note that $u'^{-1\prime}_S = \frac{1}{u''_S} < 0$ and $v'^{-1\prime} = \frac{1}{v''} < 0$. We can easily see that inequality (A.8.1) does not hold, since $v'^{-1}(\tilde{p}_{OP}) < v'^{-1}(\tilde{p}_{EPR})$ and $u'^{-1}_S(\hat{p}_{OP} - \tilde{p}_{OP}) > u'^{-1}_S(\hat{p}_{EPR} - \tilde{p}_{EPR})$. This is a contradiction. Therefore, $\tilde{p}_{OP} < \tilde{p}_{EPR}$ holds, if

$$\tilde{p}_{OP} + q > \tilde{p}_{EPR} - \hat{p}_{EPR} + P = \tilde{p}_{EPR} + p_s.$$

However, this is also a contradiction, since by our supposition $q < p_s$. Therefore, we can conclude that

$$\tilde{p}_{OP} + q < \tilde{p}_{EPR} - \hat{p}_{EPR} + P. \qquad \text{(A.8.2)}$$

If inequality (A.8.2) holds, the l.h.s. of (8.9) is smaller than that of (8.19). Then the following inequality holds:

$$v'^{-1}(\tilde{p}_{OP}) - u'^{-1}_{s}(\hat{p}_{OP} - \tilde{p}_{OP}) < v'^{-1}(\tilde{p}_{EPR}) - u'^{-1}_{s}(\hat{p}_{EPR} - \tilde{p}_{EPR}). \qquad \text{(A.8.3)}$$

Suppose that $\tilde{p}_{OP} < \tilde{p}_{EPR}$. Similarly, from Lemma 8.1, $\hat{p}_{EPR} - \tilde{p}_{EPR} < \hat{p}_{OP} - \tilde{p}_{OP}$. This supposition, however, violates inequality (A.8.3). This is a contradiction. Therefore, we obtain the desired results:

$$\tilde{p}_{OP} > \tilde{p}_{EPR} \text{ and } \hat{p}_{OP} > \hat{p}_{EPR}.$$

A.8.2 *Proof of Lemma 8.2*

Totally differentiating (8.9) and (8.10), we obtain

$$\begin{pmatrix} \frac{1}{d''} - \frac{1}{v''} - \frac{1}{u''_S} & \frac{1}{u''_S} \\ -\frac{C''_S}{d''} + \frac{C''_S}{v''} & \frac{C''_S}{u''_N} - 1 - \frac{C''_S}{C''_N} \end{pmatrix} \begin{pmatrix} \frac{d\tilde{p}}{dE} \\ \frac{d\hat{p}}{dE} \end{pmatrix} = \begin{pmatrix} 0 \\ -1 \end{pmatrix}. \qquad \text{(A.8.4)}$$

Through equation (A.8.4), we obtain the following results:

$$\frac{d\tilde{p}}{dE} = \frac{1}{u''_S|A|} > 0 \qquad \text{(A.8.5)}$$

$$\frac{d\hat{p}}{dE} = \frac{1}{|A|}\left(-\frac{1}{d''} + \frac{1}{v''} + \frac{1}{u''_S}\right) > 0 \qquad \text{(A.8.6)}$$

where $|A| = \left(\frac{1}{d''} - \frac{1}{v''} - \frac{1}{u''_S}\right)\left(\frac{C''_S}{u''_N} - 1 - \frac{C''_S}{C''_N}\right) - \frac{1}{u''_S}\left(-\frac{C''_S}{d''} + \frac{C''_S}{v''}\right) < 0.$

The total effect of E on x^*, x^*_u, and x^*_s can be analyzed by means of the following equations:

$$\frac{dx^*}{dE} = \frac{dx^*}{d\tilde{p}}\frac{d\tilde{p}}{dE} + \frac{dx^*}{d\hat{p}}\frac{d\hat{p}}{dE} \qquad \text{(A.8.7)}$$

$$\frac{dx^*_u}{dE} = \frac{dx^*_u}{d\tilde{p}}\frac{d\tilde{p}}{dE} + \frac{dx^*_u}{d\hat{p}}\frac{d\hat{p}}{dE} \qquad \text{(A.8.8)}$$

$$\frac{dx^*_s}{dE} = \frac{dx^*_s}{d\tilde{p}}\frac{d\tilde{p}}{dE} + \frac{dx^*_s}{d\hat{p}}\frac{d\hat{p}}{dE}. \qquad \text{(A.8.9)}$$

By differentiating (8.3)', (8.4)', and (8.5)', we obtain

$$\frac{dx^*}{d\tilde{p}} = 0, \frac{dx^*}{d\hat{p}} = \frac{1}{u''_N}, \frac{dx^*_u}{d\tilde{p}} = \frac{1}{d''}, \frac{dx^*_u}{d\hat{p}} = 0, \frac{dx^*_s}{d\tilde{p}} = 0, \text{ and } \frac{dx^*_s}{d\hat{p}} = \frac{1}{C''_N}.$$

Substituting equations (A.8.5) and (A.8.6) into (A.8.7), (A.8.8), and (A.8.9), the following results are obtained:

$$\frac{dx^*}{dE} = \frac{1}{u''_N|A|}\left(-\frac{1}{d''} + \frac{1}{v''} + \frac{1}{u''_S}\right) < 0 \tag{A.8.10}$$

$$\frac{dx^*_u}{dE} = \frac{1}{d''u''_S|A|} > 0 \tag{A.8.11}$$

$$\frac{dx^*_s}{dE} = \frac{1}{C''_N|A|}(-\frac{1}{d'} + \frac{1}{v''} + \frac{1}{u''_S}) > 0. \tag{A.8.12}$$

From (A.8.10) to (A.8.12) we can easily see the following result.

$$\frac{d(x^* - x^*_u - x^*_s)}{dE} = \frac{d(y^*_s - y^* - y^*_u)}{dE} < 0. \tag{A.8.13}$$

From equations (8.6)', (A.8.5), and (A.8.6) we can show that

$$\frac{dy^*}{dE} = \frac{1}{u''_S}\left(\frac{d\hat{p}}{dE} - \frac{d\tilde{p}}{dE}\right) < 0. \tag{A.8.14}$$

From equation (8.8)', we can obtain

$$\frac{dy^*_s}{dE} = \frac{1}{C''_S}(\frac{\partial\hat{p}}{\partial E} - 1).$$

Substituting (A.8.6) into this last equation, we can easily see that $dy^*_s/dE < 0$.

9 The effectiveness of the Basel Convention

9.1 Introduction

The material flow of E-waste is affected by the policies adopted in both exporting countries and importing countries. In the preceding chapter we saw how international trade, the environment, and social welfare are influenced by the introduction of an EPR policy in exporting countries.

In this chapter we shall investigate how policies adopted by importing countries affect the material flow of E-waste. The analysis is based on observations in our field studies rather than on a theoretical model. Developing countries adopt a variety of policies with regard to the importation of E-waste. Even two countries in one and the same region may have different regulations or policies regarding the importation of E-waste. Also, policies are complicated by the existence of exemptions that are sometimes intentional and sometimes not. All of them affect the material flow, as we shall see in this chapter.

The policies of importing countries are based on the Basel Convention. Note that developing countries are permitted to regulate the importation of E-waste by the Convention. So our first step is to begin with the Convention itself.

9.2 The Basel Convention

In the 1970s and 1980s a number of developed countries exported hazardous wastes to developing countries for final disposal and their disposal caused serious environmental pollution. To remedy this problem and prevent the economically motivated dumping of waste in poorer countries by rich countries, the Basel Convention came into effect in 1992. First of all, the Convention called for national self-sufficiency in hazardous waste management and for the overall minimization of hazardous waste generation and of transboundary movements of such wastes (Article 4.2 (a) and (b)). The Convention admits the right of parties to the Convention to prohibit the import of hazardous wastes and prohibits parties from exporting hazardous wastes to those parties that have prohibited the import of such wastes (Article 4.1). The Basel Convention requires that prior notice of any proposed export of hazardous wastes be given to the government of an importing country and its importation be approved by that government

(Article 6.3 and 4). In addition, the Convention prohibits parties from trading hazardous wastes with non-parties (Article 4.5).

It must be stressed that the Convention does not unconditionally prohibit OECD countries from exporting hazardous wastes to non-OECD countries but admits the right of the latter to prohibit the importation of hazardous wastes. In this sense, the Convention is an international trade policy that is led by importing countries.

The hazardous wastes to be regulated under the Convention are clarified in List A of Annex VIII. The list includes waste having as constituents or contaminants, excluding metal waste in massive form, any of the following: antimony, beryllium, cadmium, lead, selenium, and tellurium. Therefore, "mixed metals" (including E-waste scrap) that are to be exported from developed to developing countries for recycling must be regulated by the Convention.

The Convention also provides a list of unregulated wastes or items (Annex IX List B). This list includes any mixed non-ferrous metal that does not contain hazardous substances such as lead, mercury, cadmium, antimony, and so on (B1050). Electrical and electronic assemblies (including PCBs (printed circuit boards), electronic components, and wires) are also included in the same list if they are destined for direct reuse, and not for recycling or final disposal (B1110). Accordingly, second-hand items and electronic components that are to be reused without any further processing are not regulated by the Convention.

9.3 Policies of developing countries with respect to second-hand items and mixed metal scrap

Despite the Basel Convention, the transboundary movement of hazardous waste, especially of E-waste, has increased rather than decreased, as we saw in Chapter 7. E-waste is exported from developed countries to developing countries as second-hand items and mixed metal or E-waste scrap. The former is not regulated by the Convention. In addition, the latter is also unregulated if hazardous substances are not included. When the proportion of hazardous substance content is sufficiently low (and usually it is not high), the importation of mixed metal scrap might be tacitly permitted. In this sense, the regulations on E-waste (including second-hand items) are not effective.

Thus it is possible to get around the Convention, with the result that it does not control the material flow of E-waste. In response to this lax regulation by the Convention, some developing countries have implemented more restrictive regulations on the importation of E-waste. Tables 9.1 and 9.2 show individual regulations in the Asian region on the trade of mixed metal scrap (or E-waste scrap) and second-hand items, respectively.

As we see in these tables, Asian countries have individual regulations on the importation of E-waste scrap and second-hand items that are even stricter than the Basel Convention. Though these regulations exist on paper, however, in actual practice their enforcement is weak. For example, while China bans the importation of E-waste scrap, China imports a huge amount of E-waste scrap for

Table 9.1 Regulations on the importation of mixed metal scrap (E-waste scrap) in Asia

Country	*Import regulation on mixed metal scrap*	*Transit regulation*
China	Ban	Ban
Taiwan	Permission required	Permission required
Korea	Ban	Permission required
Hong Kong	Permission required	Ban (applied for OECD countries)
Vietnam	Ban	Ban
Cambodia	Ban	No regulation
Thailand	Permission required	Permission required
Philippines	Ban	Permission required
Malaysia	Permission required	Permission required
Singapore	Permission required	Permission required
Indonesia	Permission required	Permission required
India	Permission required	Permission required

Source: Table 2.11.1 in Terazono *et al.* (2009).

Note
Although China bans the importation of mixed metal scrap, authorized (or licensed) firms can import or export it.

Table 9.2 Regulations on the importation of second-hand items in Asia

Country	*Import regulation on second-hand items*	*Conditions required*
China	Ban	
Taiwan	NA	
Korea	No regulation	
Hong Kong	Conditional	Items that were produced no later than five years earlier and can be directly used without any repairs can be imported.
Vietnam	Basically a ban	Second-hand PCs and home appliances other than laptop PCs cannot be imported.
Cambodia	No regulation	As an exception, the importation of PCs is banned.
Thailand	Conditional	Items that were produced no later than three years earlier can be imported.
Philippines	Permission required	
Malaysia	No regulation	
Singapore	No regulation	
Indonesia	Ban	
India	Ban	

Source: Table 2.11.3 in Terazono *et al.* (2009).

recycling from Japan, the EU, and the US, as shown in Table 7.5 in Section 7.4. Tables 7.9 and 7.10 in Section 7.4 also show that a huge amount of second-hand items are exported to Vietnam and China, regardless of the fact that both these countries ban the importation of these items. There are several reasons for the weak enforcement of the law.

Foremost among the reasons is, as stated in Section 7.3, that several developing countries, including China and Vietnam, have in recent years experienced rapid economic growth, with the result that the demand for mixed metal scrap and second-hand electrical and electronic equipment (EEE) has dramatically increased. Per capita incomes also have increased, but they have not reached a sufficiently high level so that consumers can afford to purchase new products. Hence many developing country governments inevitably experience enormous economic pressure to import E-waste scrap and second-hand items.

Another reason is that corrupt agents of governments in developing countries relax the enforcement of their country's regulations on the importation of E-waste scrap and second-hand items. Many Asian countries require permission with respect to the importation of E-waste scrap (see Table 9.1). As Rose-Ackerman (2006) points out, the allocation of import and export permissions is a frequent source of payoffs and patronage with bribes.[1] As for second-hand items, it is easier to smuggle them into a country because they are so highly valued and so compact in shape.

A third reason is that, in several developing countries, intentionally or not, there are loopholes or exemptions in the regulations. The government of China tacitly permits the importation of second-hand items as long as they are rebuilt and then reexported. In India, the importation of second-hand items is also permitted when they are destined for donation to educational institutions.

The next section provides several examples of material flows of second-hand items and E-waste scrap in the form of PCBs. We can see connections between the material flows and the policies adopted in importing countries. The first example is of the material flows revolving around China, Vietnam, and Cambodia, and the second one is of those around India.

9.4 Relations between policies and the material flows of E-waste around China[2]

In this section we investigate the material flow of second-hand EEE and E-waste scrap, with emphasis on China, Vietnam, and Cambodia. We also focus on PCBs among E-waste scrap. Remember that China and Vietnam have banned the import of second-hand EEE and PCBs. Yet in both countries a considerable amount of second-hand EEE and PCBs are traded. The reason is the large gap between policies in principle and implementation in practice.

9.4.1 The material flows of second-hand EEE

First, let us look at the trade of second-hand EEE between China and northern Vietnam. In principle, China enforces an absolute ban on the import of second-hand EEE. But in practice large amounts of second-hand EEE are imported into China. This contradiction is attributed to the fact that the government of China allows second-hand EEE to be imported as long as the items are rebuilt and then reexported. Indeed, the government of China encourages the export of second-hand products by reducing the value-added tax (VAT) on second-hand products to 4 percent. Although the import of second-hand EEE for the purpose of reexport was banned after 2002, in actuality this type of import activity is still unofficially allowed to go on (Yoshida, 2006).

The domestic demand in China for second-hand home appliances, especially television sets, is decreasing because such items have to compete in price with new domestically made products, whereas the demand for second-hand home appliances in other developing countries is still very high. For that reason, China has imported second-hand home appliances (especially second-hand cathode-ray tube—or CRT—television sets) from Japan and the US, rebuilt or renewed them (mainly in Guangzhou), and then reexported them to other countries, including Vietnam. There are two import routes for this activity. In the first, second-hand EEE is imported via Hong Kong from Japan and the US and then brought to Guangzhou for rebuilding. The rebuilt second-hand EEE is eventually smuggled into Vietnam via the Dongxin border area.

The second route is roundabout. First of all the CRT television sets are imported at Haiphong international port (in northern Vietnam) from Japan and the US, from which they are exported (from Vietnam) through the Mong Cai border gate[3] to Dongxin in China. They are then taken to Guangzhou by truck, rebuilt in Guangzhou, and finally exported illegally back to Vietnam. Note that Vietnam also bans the import of second-hand EEE in principle (see Table 9.2). However, the Vietnamese government applies the same exemption to the import of second-hand EEE that China does; that is, it allows the importation of second-hand EEE for the purpose of reexport. This makes it possible for China to import second-hand EEE from Vietnam by this route.

The smuggling between Dongxin and Mong Cai is quite simple. Second-hand EEE is carried over the river in small boats, covered only with blue canvas. Second-hand home appliances smuggled from Dongxin into Mong Cai have been put up for sale in second-hand markets in northern Vietnam, including in Hanoi and Haiphong.

Not all the second-hand EEE imported into China, however, is eventually exported to other countries. In China there is still a large demand for second-hand computers, even though the demand for home appliances has been decreasing. Chinese business firms in particular need a great deal of second-hand computers. The second-hand computers imported from Vietnam ("for the purpose of reexport") remain in China and are sold in the second-hand markets. Customs officials in Dongxin tacitly permit the importation of second-hand computers even though they know that the goods will remain in China.

Vietnam also allows second-hand EEE to be imported for the purpose of reexport, despite the fact that not all of them are in fact reexported. Some of them (especially second-hand home appliances) remain in the country and are sold in the domestic markets. It seems difficult for both governments to check that all the second-hand EEE imported for reexport is really exported.

On the basis of what we observed, we are inclined to conclude that the trade of second-hand EEE between China and Vietnam is being carried out with at least partial consent by the two governments.

The smuggling of second-hand EEE in the south of Vietnam is a little different from that in the north. In the south, a certain amount of second-hand EEE has been illicitly imported, mainly from Japan and the US. Since May 2006, however, the ban on importing such items has been tightened, and almost all the second-hand EEE sold in the south is now smuggled in from Cambodia, where control over such items is much laxer.

Cambodia imports second-hand EEE through Sihanoukville international port, then transports it to Phnom Penh by truck. Containers of imported second-hand EEE are put up for auction every weekend. A number of Vietnamese dealers also participate in these auction sales. Some of this second-hand EEE is smuggled into Vietnam through the border areas in the south between Vietnam and Cambodia. There are up to seven hot-spot smuggling places; huge amounts of second-hand EEE are smuggled especially between Go Tamau (Cambodia) and Chau Doc (Vietnam), and between Babet (Cambodia) and Moc Bai (Vietnam). A portion of the smuggled EEE is consumed in the Mekong River delta in southern Vietnam, while the remainder is transported to Ho Chi Minh City for sale. The Nhat Tao market in Ho Chi Minh is a central market for second-hand EEE. In Ho Chi Minh one rarely sees original imported television sets for sale. In most cases, the glass of the CRTs has been polished and the original covers and PCBs have been replaced by new ones.

Not all the second-hand EEE imported into Cambodia is smuggled into Vietnam. On Kampuchea Krom Street in Phnom Penh there are about 150 shops selling second-hand EEE. The country also has a large-scale second-hand market. Cambodia allows the import of any second-hand EEE except second-hand computers (see Table 9.2), yet there are still nearly 20 shops selling second-hand computers on Kampuchea Krom Street in the capital. Most of the time these second-hand computers are concealed at the bottom of containers being smuggled into Cambodia. Customs officials in Sihanoukville are aware of the methods used in illicit operations, but they still stop short of strict regulation.

The prices of second-hand television sets and computers in Cambodia, Vietnam, and China are shown in Table 9.3. We can see that the price of a second-hand TV set is much higher in Vietnam than in Cambodia and China, which explains why second-hand items like TV sets have been flowing from

Table 9.3 Selling price of a second-hand television set and a computer set

	Phnom Penh (Cambodia)	*Ho Chi Minh (Vietnam)*	*Beijing (China)*
A second-hand personal computer set (including a monitor)	\$40~ 100	\$ 70~ 120	\$120~ 220
Second-hand (21-inch) television set	\$50	\$70	\$40

Source: This table is a reprint of Table 1 in Shinkuma and Huong (2009).

Cambodia and China to Vietnam. On the other hand, the price of a second-hand computer is highest in China, followed by Vietnam, so this explains why second-hand computers are exported to China from Vietnam and from Cambodia to Vietnam.

With all the data taken into account, the flow of second-hand EEE can be illustrated as shown in Figure 9.1. Note that there are two routes in northern Vietnam, one represented by solid lines and the other by broken lines.

As shown in Tables 7.9 and 7.10 in Section 7.4, second-hand CRT television sets and computer monitors have been exported from Japan mainly to Hong Kong. Note that Hong Kong is just a transition point in the trade and the second-hand EEE imported there gets reexported to other countries, particularly China. According to Table 7.9, the number of CRT television sets exported to Vietnam is relatively small. On the other hand, the numbers exported to China and Cambodia are relatively large. This statistical data does not contradict the material flow described earlier in this section, since second-hand TV sets are reexported (or smuggled) into Vietnam from China and Cambodia. The statistical data shown in Table 7.10 indicates that the number of computer monitors imported into Vietnam is larger than the number imported into China, while the demand for second-hand computers is lower in Vietnam than it is in China. These statistics also do not contradict our observations, since second-hand computers and monitors get reexported from Vietnam into China.

It is quite difficult to grasp the actual flow of second-hand EEE by looking only at the trade statistics. Smuggling generates large gaps between statistical data and facts.

9.4.2 The material flow of PCBs

In those countries that import second-hand EEE, many repair and maintenance shops can be found in the second-hand markets. Workers in such repair shops replace damaged or worn-out components with new components made in China. Any components that are seriously damaged and incapable of being reused still have a positive sales value because they can be recycled. A PCB in particular is most valuable. PCBs in Cambodia that are damaged beyond reuse are exported illegally to Vietnam. Of these, only a small portion is recycled in Ho Chi Minh City; most of them are transported to northern Vietnam, from which they are illegally exported to China via Mong Cai.

Guangdong Province in China is the largest recycling center of PCBs in the world; the area around Guiyu in the same province is the most famous, as we saw in Section 7.2. But because the Chinese government has strengthened controls over the importing of PCBs, recently several Chinese firms have begun to recycle PCBs in Mong Cai in Vietnam.

National policies also have strong effects on the flow of PCBs. In principle, China bans the import of PCBs, especially from developed countries, and yet unofficially it has allowed the import of PCBs from Vietnam. Taking advantage of this loophole, some dealers have imported E-waste scrap, including PCBs,

Figure 9.1 Material flow of second-hand EEE around Vietnam (source: this table is a reprint of Figure 1 in Shinkuma and Huong (2009)).

from Japan and the US into Haiphong and then dismantled the items in areas surrounding Haiphong before exporting the scrap to China through Mong Cai. To remedy the situation, the government of Vietnam has, since October 2004, banned the import of E-waste scrap for the purpose of reexport (it is not clear whether or not this was in response to requests made in this regard by the government of China). Additionally, at present both the importing and dismantling of E-waste scrap are strictly prohibited in Vietnam. As a result of this policy, the amount of imported E-waste scrap, including PCBs, has decreased dramatically.

It is assumed, therefore, that the PCBs exported from Vietnam to China have their origin in Cambodia and Vietnam.

Although China bans the import of PCBs from Japan in principle, a great deal of PCBs generated in home-appliance recycling factories in Japan have been imported into China along with other metal scrap (as a form of mixed metal). In practice, if the amount of PCBs in a consignment of mixed metal scrap does not exceed 3 percent of the total amount, the entire consignment can be allowed into China.

Table 9.4 shows the buying prices of PCBs in southern Vietnam, northern Vietnam, and Guangdong Province in China. The price is lowest in the south of Vietnam, higher in the north, and highest in Guangdong Province. These figures support our observations regarding the flow of PCBs. Table 9.4 also indicates the buying price offered by a firm in Hangzhou, China, that does proper recycling of PCBs with the authorization of the Chinese government. The price is quite low, which indicates that it is costly to minimize the adverse environmental effects. Because of the low price, this recycling firm cannot collect enough PCBs to make a profit. The recycling method adopted in Guangdong Province does not take into account the benefit of counteracting environmental effects; if it were to take such a benefit into account, the firm would be able to buy PCBs at a higher price.

9.5 Relations between the policies and the material flows of E-waste around India

Since the 1990s India has been another center of E-waste recycling. According to the Central Pollution Control Board (CPCB), every year 146,000 tons of E-waste are recycled in India. The GTZ (Deutsche Gesellschaft für Technische Zusammenarbeit), however, estimated the total amount of E-waste in India was around 330,000 tons in 2007.[4]

E-waste is recycled in Delhi, several states around Delhi, Chennai, Bangalore, Pune, and Kolkata, with Delhi acting as the center of E-waste recycling in the whole country. In the Mandoli Industrial Area in Delhi, Muslim residents engage in illegal recycling of PCBs using acid bath processes to recover copper and gold.

India ratified the Basel Convention in 1992. The government of India has banned the import of second-hand EEE (see Table 9.2). According to the GTZ, however, 50,000 tons of E-waste were illegally imported into the country in 2007.[5] The illegal importation is due to a loophole in the law. In India it is

Table 9.4 Buying prices of PCBs per ton in China and Vietnam

Chau Doc (the south of Vietnam)	\$475
Trang Minh village (the north of Vietnam)	\$600
Guiyu town (Guangdong Province, China)	\$420 ~ \$1,900
An authorized recycling firm (Hangzhou, China)	\$250

Source: This table is a reprint of Table 4 in Shinkuma and Huong (2009).

permitted to import second-hand EEE, including PCs, only for the purpose of donating them to educational institutions, and no tariff is imposed on second-hand EEE imported for that purpose. In practice, though, only a small proportion of second-hand EEE has been donated to educational institutions, with most of it being sold in used-goods markets.

It is a feature to be emphasized that in India E-wastes are imported as second-hand items or parts even if they are recycled. Second-hand items or parts such as PC motherboards are imported, either directly or via the UAE, from EU member countries, the US, Japan, Singapore, Malaysia, and Korea. The amount of items imported from the EU and the US via the UAE is particularly large. Just as Hong Kong is a transit point of E-waste, so Sharja near Dubai, the largest city in the UAE, is the primary distribution center between the EU and the eastern half of the US, and developing countries in South Asia, the Middle East, and eastern Africa.

Sharja, UAE

The second-hand market in Sharja is a compound in which is located a huge number of small shops, but shops dealing with the same categories, such as PCs or home appliances, are located close to one another, and the market is divided according to items, as shown in Figure 9.2. A large proportion of the second-hand PCs are imported from EU member countries (especially the UK) and from the US. Among other home appliances, such as TV sets, washing machines, and air conditioners, we can find second-hand items imported from Asian countries, especially Japan, Singapore, and Korea. Many dealers visiting the marketplace come from South Asia (mainly India and Pakistan), the Middle East (mainly Iran and Saudi Arabia), and Africa (mainly Egypt, Tanzania, Uganda, and South Africa).

Recently, in India huge amounts of second-hand PC motherboards have been illegally imported from China. Some of them are sold in second-hand-goods markets and the others are recycled in the Mandoli Industrial Area in Delhi. There are two major routes for the illegal importation from China. The first route is by ship from China to Delhi via Mumbai, Kolkata, and Chennai. The second one is by train and truck from China to Delhi via Kathmandu in Nepal. The first route is used by those who take advantage of the donation-to-education regulation; the second one is used by out-and-out smugglers.

The above information is based on interviews with several scrap dealers in Delhi and Mumbai and with government officials in Nepal. Table 9.5 shows the buying prices of PCBs in the informal sectors of China and India. The buying price in India is much higher than that in China. The data in the table show that it is profitable to import PCBs from China. The fact that E-waste is being imported into India from China may alter the global picture of E-waste material flows. In an earlier chapter we concluded that China and India have been importing E-waste from separate areas: China from Japan, Korea, Southeast Asia, and the western half of the US; and India from the EU, central East Asia, Southeast Asia and the eastern half of the US. India's importation of PCBs from China

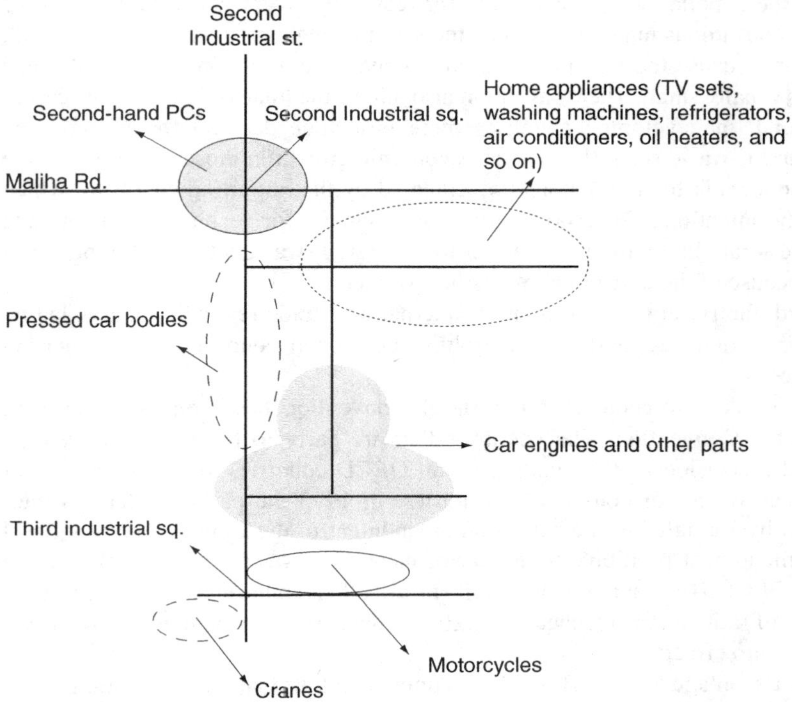

Figure 9.2 The second-hand market in Sharja, UAE (source: Field survey in Sharja from November 1–5, 2009 (by Shinkuma with Keisaku Higashida)).

Table 9.5 Buying prices of FCBs per ton in China and India

Informal sector in China (Guiyu town, Guangdong Province)	\$420 – \$1,900
An authorized recycling firm in China (Hangzhou)	\$250
Informal sector in India (Chennai)	\$3,000 – \$3,600
An authorized recycling firm in India (Chennai)	\$750 – \$1,200

Source: This table is a reprint of Table 1 in Shinkuma and Managi (2010a).

suggests that in the near future India might become a larger center of E-waste recycling than even China.

9.6 Weak enforcement under the Basel Convention

We can now summarize the material flows of E-waste under the Basel Convention. Enforcement is unfortunately weak, and the environment has been seriously polluted, especially in China, India, and Pakistan. There seem to be several reasons why regulation is not effective under the Convention. First, the Convention itself allows the exportation of metal scrap for recycling except when the metal scrap includes hazardous materials. Enforcement of this regulation tends to be difficult, however, because the amount of hazardous materials in a consignment of metal scrap is usually small. The Convention also allows the trade of E-waste for reuse.

Second, in developing countries there is a huge demand for second-hand items and E-waste scrap due to rapid economic growth in those countries. There are loopholes in the regulations implemented by the importing countries, which might be intentional in response to a huge demand for second-hand items and E-waste scrap. In addition, it is easier to smuggle E-waste into developing countries because of the corruption of customs officers.

Third, the Basel Convention is an international trade regulation that is led by importing countries. It does not prohibit developed countries from exporting E-waste.

The weak enforcement of the Basel Convention has prompted moves to amend the Convention. The countries that are party to the Basel Convention adopted a decision in 1994 calling on all OECD countries to ban the export of hazardous wastes to non-OECD countries. In 1995 the parties reiterated their concern by formalizing the ban as an amendment to the Convention; this Basel Ban Amendment prohibits the export of hazardous wastes from member states of the OECD, EU, and Liechtenstein to all other countries. Though the total number of ratifications reached 69 in 2010, however, the ban amendment has yet to come into effect.

The advantage of the Basel Ban Amendment is that the regulation is by exporting countries. As seen in this chapter, the regulations by importing countries are not effective. Economic pressures to export E-waste are not as strong as the economic pressures to import it. The level of compliance with the law is generally higher in developed countries. So it is much easier to regulate E-waste trade from the exporting side.

Should we, then, cooperate with one another and make the amendment come into force? Before we answer that question, we ought to examine the policies for discouraging improper recycling that have been adopted in the centers of E-waste recycling. If these policies have great potential for discouraging improper recycling, the OECD member countries could support such policies institutionally, technically, and financially rather than ban the export of E-waste. Then we could take advantage of the current situation, in which E-waste recycling processes are concentrated in certain regions. So in the next chapter we shall examine the effectiveness of the policies adopted in China and India for responsible recycling of E-waste.

10 The policies adopted in China and India and the international trade policy

10.1 Introduction

In this chapter we look at the effectiveness of the general policy for discouraging improper recycling that has been adopted in China and India: a licensing system for E-waste recycling (Section 10.2). We then evaluate the Basel Ban Amendment by weighing the advantages of such a policy against its disadvantages (Sections 10.3 and 10.4). Finally, we propose an alternative trading regulation system, one that is an extended license system that involves not only importing countries but also exporting countries (Sections 10.5 and 10.6).

10.2 The policies adopted in China and India[1]

10.2.1 China

A huge amount of E-waste has been imported from developed countries and recycled in China's coastal areas since 1992. Improper recycling of E-waste, particularly PCBs (printed circuit boards), by the informal sector has generated serious environmental pollution. Guiyu in Guangdong Province has been the center of PCB recycling.

At first the Chinese government banned improper recycling by the informal sector when the country was confronted with the magnitude of environmental pollution. However, regulations that necessitated detection of offenders failed because recycling facilities in the informal sector tended to be relatively small, and thus easy to conceal or to shift to other areas. The ban on improper recycling only resulted in the spread of improper recycling activities to other places.

Hence, since the late 1990s the Chinese government has changed its policy from banning improper recycling to encouraging proper recycling, through the creation of proper recycling areas (recycling industrial parks),[2] and through attracting to them selected firms. It introduced a license scheme for E-waste recycling, and recycling by unauthorized recyclers has been prohibited by law (Regulation for Preventing Environmental Pollution Caused by WEEE). In addition, the disposers of E-waste are obliged to sell the waste to licensed recyclers. Table 10.1 summarizes the licensing scheme adopted in China. The structure of

Table 10.1 The Regulation for Preventing Environmental Pollution Caused by WEEE (2007)

Article 5	Disassembly and recycling of WEEE requires a license.
Article 7	List of licensed recyclers or WMFs is to be published. Improper treatment including illegal disposal is prohibited. Reconsignment of WEEE to non-license holders is also prohibited. Operation without license is prohibited. If a license holder is detected violating the regulation twice or more than twice in three years, the license is to be suspended.
Article 8	Any trade and processing should be recorded, including what kind of and how much WEEE is traded, the names of the trading partners, and how the goods are treated.
Article 14	Producers and retailers of EEE are obliged to collect end-of-life EEE.
Article 15	Disposers of end-of-life EEE are required to consign it to licensed recyclers or WMFs.
Article 16	Organizations that dispose of WEEE should record what kind of and how much WEEE is consigned to licensed WMFs, including the names of those WMFs.
Article 20	If operation without license is detected, a fine (between 50,000 CNY and 500,000 CNY) is imposed.
Article 21	If reconsignment of WEEE to non-license holders is detected, a fine (no more than 30,000 CNY) is imposed.
Article 23	If a license holder is detected polluting the environment, the license is suspended and the period of suspension depends on the seriousness of the pollution.

Note
1US$ = 6.78CNY (September 8, 2010).

the regulation is quite similar to the license scheme that has been adopted for the control of industrial waste in developed countries (see Chapter 2). To enhance the disposers' responsibility in the international trade of E-waste, the government of China not only allocates the rights to import E-waste to authorized (or licensed) recycling firms in recycling industrial parks but also allocates the rights to export E-waste scrap to Chinese-funded firms in the countries that originally exported the E-waste, such as Japan and the US (Terazono *et al.*, 2007).

10.2.2 India

Since the 1990s India has been another center of E-waste recycling and, as in China, the environment has been seriously polluted by improper recycling (see Chapter 7 and also Section 9.5). The government of India also banned such environmentally disastrous recycling. Again it was hard to control and enforce the ban because such recycling facilities are usually managed by households or small firms, so they can easily be shifted to other areas. Mandatory removal only resulted in improper recycling activities spreading to other places.

When a government is confronted with ineffective enforcement, it seems to change its policy from simple prohibition of improper recycling to the promotion of proper recycling. The Indian government, through its Central Pollution Control Board (CPCB), issues licenses to proper recycling firms. In 2009 12 recycling firms applied for a license and only six of them were granted one; four of them operate in Chennai, and two in Bangalore. There are no licensed firms in Delhi, Mumbai, or Pune, although six have applied (three in Delhi, two in Mumbai, and one in Pune). Thus the policy response in India is similar to that made by the Chinese government, although China is ahead of India in terms of the capacity of its authorized recycling firms.

10.2.3 License scheme

In both countries the enforcement policy has changed from simple prohibition of improper recycling facilities to promotion of proper recycling through the introduction of a license scheme. In China, to promote proper recycling, the law regulating E-waste recycling obliges disposers of E-waste to sell it to licensed recycling firms. The government of China allows only authorized firms to import and export E-waste, issuing them rights to both import and export. As a result, a huge amount of imported E-waste scrap is recycled in the recycling industrial parks.

In India, on the other hand, there is no onus on disposers, or at least it is lighter than that in China. A licensed firm in Chennai in India, however, can buy a sufficient amount of E-waste, even though the buying price of PCBs is less than half the price offered by the informal sector in the same city (see Table 9.5). Several major IT firms (IBM, INTEL, and ACER) sell end-of-life PCs and motherboards at a production loss to the licensed firm in spite of the low selling price. They do not want to risk hurting their reputations if any contact with the informal sector is detected and made known to the public.

In the next section we use an economic model to assess the effectiveness of a license scheme.

10.3 The effectiveness of a license scheme[3]

Let us use a simple theoretical model to assess the effectiveness of a license scheme for E-waste recycling. Our model is a variant of the state-dependent regulation model that was first developed by Harrington (1988).

Recyclers (including those in the informal sector) buy E-waste from the disposers (the buying price per ton is denoted by p),[4] recover metals from it, and sell the metals to smelters for a market price per ton denoted by p_m, which is assumed to be constant.

Suppose that the government sets a standard for proper recycling of E-waste. The environmentally sound technology meeting the standard (including a suitable recycling facility) generates an additional cost. For the sake of simplicity, the cost of recovering resources from E-waste is assumed to be zero. The additional cost for properly recycling one ton of E-waste is denoted by θ. We assume that θ

differs among recyclers (including the informal sector) and θ distributes between 0 and $\bar{\theta}$. This assumption relies on the fact that recyclers differ from one another in respect to their knowledge of proper recycling technology and its financial accessibility. Again, for the sake of simplicity it is assumed that the capacity of each recycler is fixed at 1 ton.

We assume that the government introduces a license scheme characterized as follows. A license for proper E-waste recycling is issued to any recycler (including one from the informal sector) who applies for the license.[5] Both improper recycling and operations by unlicensed recyclers are prohibited by the law and the regulator implements a periodic inspection. The probability of improper recycling and operations by unlicensed recyclers being detected is denoted by μ.[6] If improper recycling or operating without a license is detected by the regulator, a fine denoted by f is imposed on the recycler. In addition, if the recycler caught doing improper recycling is a license holder, that license must be suspended for the next period. At the end of that period, the suspended license is returned to the recycler provided the recycler is not again caught operating without a license during that same period. Table 10.2 sketches how the transition of recyclers' status depends on their actions and on whether or not they are detected by the regulator.

In addition, we assume that the law prohibits disposers of E-waste from selling it to unlicensed recyclers. If a transaction with an unlicensed recycler is detected, a fine denoted by f' is imposed on the disposer. The probability of detection is denoted by μ', and the expected fine imposed on the disposer is given by $\mu' f'$. In the equilibrium disposers must be indifferent as to contracting with a licensed recycler or an unlicensed recycler.[7] If the E-waste price per ton offered by a licensed and an unlicensed recycler is represented by p_1 and p_0, respectively, the following equation must be satisfied in the equilibrium:

Table 10.2 Recyclers' decisions, detection, and transition of status

Current period			*Next period*		
Status	*Action*	*Detection*	*Status*	*Action*	*Detection*
Compliant licensed recycler	Proper recycling	No →	Compliant licensed recycler	Proper Recycling	No
Noncompliant licensed recycler		Yes →	Unlicensed recycler		Yes No
	Improper Recycling	No →	Noncompliant licensed recycler	Improper Recycling	Yes No
Unlicensed recycler		Yes →	Unlicensed recycler		Yes No
		No →	Noncompliant licensed recycler		Yes No

Source: This table is a reprint of Table 2 in Shinkuma and Managi (2010a).

$$p_0 = p_1 + \mu' f'. \tag{10.1}$$

Consequently, the buying price for an unlicensed recycler (p_0) is higher than that for a licensed recycler (p_1).

Recyclers have two options: proper recycling or improper recycling. Intuitively, recyclers who are licensed and have a relatively lower value of θ, will comply with the standard for proper recycling. On the other hand, recyclers with a higher value of θ will not comply with the standard. Note that all recyclers will want to be licensed, because the buying price of E-waste is lower for a licensed recycler (see (10.1)). Therefore, in the stationary state, there are three types of recyclers: licensed recyclers who will comply with the standard (let us call them *compliant licensed recyclers*), licensed recyclers who will not comply with the standard but will still keep their licenses (let us call them *noncompliant licensed recyclers*), and recyclers whose licenses are suspended (let us call them *unlicensed recyclers*). Note that unlicensed recyclers have no incentive to properly recycle E-waste, because the operations of unlicensed recyclers are already illegal.

To obtain the steady-state equilibrium, we have to compare the current value profits of proper and improper recycling with each other. Before calculating it we need to look at the following list of notations.

π_1^p: The current value of the total amount of profit that a *compliant licensed recycler* can earn from the current period to an indefinite future period. The superscript p and the subscript 1 represent proper recycling and the licensed recycler, respectively.

π_0^i: The current value of the total amount of profit that an *unlicensed recycler* can earn from the current period to an indefinite future period. The superscript i and the subscript 0 represent improper recycling and the unlicensed recycler, respectively

π_1^i: The current value of the total amount of profit that a *noncompliant licensed recycler* can earn from the current period to an indefinite future period.

β: The discount factor[8] (constant).

First we derive the current value of profit for proper recycling. A *compliant licensed recycler* earns $p_m - p_1 - \theta$ every period. Then the following equation must be satisfied:

$$\pi_1^p = p_m - p_1 - \theta + \beta\pi_1^p. \tag{10.2}$$

We can easily calculate π_1^p as follows:

$$\pi_1^p = \frac{p_m - p_1 - \theta}{1 - \beta}. \tag{10.3}$$

Second, the current value of profit for improper recycling can be derived as follows. The profit of an *unlicensed recycler*, π^i_0, must satisfy the following equation:

$$\pi^i_0 = \underbrace{p_m - p_0 - \mu f}_{\text{The profit earned at the current period}} + \beta \underbrace{\{(1-\mu)\pi^i_1 + \mu\pi^i_0\}}_{\text{The expected profit earned from the next period to an indefinite future period}} \tag{10.4}$$

Note that the expected fine imposed on improper recycling is denoted by μf. In addition, only if the recycler is not caught operating without a license (or improperly recycling) during the current period, the suspended license is returned to the recycler at the beginning of the next period and he/she becomes a *noncompliant licensed recycler* in the same period (the probability is $1 - \mu$). Otherwise, the license of the recycler is suspended in the next period (see Table 10.2). Therefore, the expected profit earned by an *unlicensed recycler* from the next period to an indefinite future period can be written as the bracket in the second term of equation (10.4).

Similarly, as for π^i_1 (the current value of profit of a *noncompliant licensed recycler*), we can obtain the following equation:

$$\pi^i_1 = p_m - p_1 - \mu f + \beta\{(1-\mu)\pi^i_1 + \mu\pi^i_0\}. \tag{10.5}$$

Note that a *noncompliant licensed recycler* has to pay a fine (f) and loses the license in the next period if improper recycling is detected (the probability is μ), but otherwise the recycler can keep the license in the next period. Therefore, the expected profit earned by a *noncompliant licensed recycler* from the next period to an indefinite future period can be written as the bracket in equation (10.5).

By (10.4) and (10.5) we can calculate π^i_0 and π^i_1.

$$\pi^i_1 = \frac{(1-\beta\mu)(p_m - p_1 - \mu f) + \beta\mu\{p_m - p_0 - \mu f\}}{1-\beta} \tag{10.6}$$

$$\pi^i_0 = \frac{1}{1-\beta\mu}[p_m - p_0 - \mu f + \beta(1-\mu)\pi^i_1]. \tag{10.7}$$

Improper recyclers include both *noncompliant licensed recyclers* and *unlicensed recyclers*. Denote by α and $1 - \alpha$ the relative proportion of *unlicensed recyclers* and *noncompliant licensed recyclers* among all improper recyclers, respectively. In the steady state α must stay at a unique value. Note that *unlicensed recyclers* can apply for and get licenses in the next period if they are not caught operating without a license (or improperly recycling) during the current period. Then, in the steady-state equilibrium the proportion of *unlicensed recyclers* who will acquire licenses in the next period ($(1 - \mu)\alpha$) must be equal to the proportion of the *noncompliant licensed recyclers* who lose their licenses for the next period ($\mu(1 - \alpha)$). Therefore, we can obtain the steady-state value of α,

$$\alpha = \mu. \tag{10.8}$$

The expected profit of an improper recycler represented by $E(\pi^i)$ in the steady state can be calculated as

$$E(\pi^i) = \alpha\pi_0^i + (1 - \alpha)\pi_1^i. \tag{10.9}$$

As stated above, whether recyclers choose proper or improper recycling depends on the level of θ. In other words, only licensed recyclers with a lower value of θ comply with the standard for proper recycling. We are ready to derive the critical value of θ that is represented by $\hat{\theta}$. Since the marginal licensed recycler with $\hat{\theta}$ must be indifferent as to compliance (or proper recycling) or noncompliant behavior (or improper recycling), the following equation must hold:

$$\pi_1^p = E(\pi^i). \tag{10.10}$$

By substituting (10.6), (10.7), and (10.8) into (10.9) and solving (10.10), we can derive the critical value of θ (or $\hat{\theta}$):

$$\hat{\theta} = \mu f + \mu(p_0 - p_1). \tag{10.11}$$

Note that an increment of $\hat{\theta}$ implies that the proportion of compliant recyclers increases and improper recycling decreases.

There are two policy options for promoting proper recycling and discouraging improper recycling. The first option is raising the fine imposed on improper recycling (f). The first term at the right-hand side of (10.11) represents the effect of the fine on improper recycling.

The second option is licensing. Its effect is represented by the second term at the right-hand side of (10.11). To see this, suppose that licensing is not available and the regulator can only inspect and impose a fine on improper recyclers. In this setting, note that the market price must be unique ($p_0 = p_1 = p$). The profits from proper recycling and from improper recycling can be calculated respectively as follows:

$$\pi^p = \frac{P_m - p - \theta}{1 - \beta}.$$

$$\pi^i = \frac{P_m - p - \mu f}{1 - \beta}.$$

Noting that the critical value $\hat{\theta}$ must satisfy $\pi^p = \pi^i$, we can calculate it explicitly.

$$\hat{\theta} = \mu f. \tag{10.12}$$

Note that the right-hand side of (10.12) represents the effect of a fine on improper recycling. This shows that the second term at the right-hand side of (10.11) represents the original effect of licensing. So we can state that a license scheme is effective in promoting proper recycling and in discouraging improper recycling.

It must be stressed that a license scheme can discourage improper recycling even if the government does not impose any fine on improper recycling, i.e., if $f = 0$. How can a license scheme discourage improper recycling without any fine on improper recycling? Such a result follows from the basic idea behind a license scheme, namely, that a compliant license holder can be subsidized ($p_0 - p_1 > 0$), and consequently his profit must be larger than that of an unlicensed recycler. The suspension of the license plays a role similar to that of a fine. Some licensed recyclers prefer compliance with the standard because the potential loss from improper recycling is larger than the potential benefit.

Equation (10.11) does not mean, however, that a license scheme can always function effectively. Effectiveness depends on the disposer's responsibility. This can be seen if we consider a special case. Suppose that the law does not prohibit disposers of E-waste from selling to unlicensed recyclers. This implies that $f' = 0$. It follows from equation (10.1) that price discrimination between a licensed recycler and an unlicensed recycler is eliminated, i.e., $p_0 - p_1 = 0$. In this case, a license scheme becomes ineffective, as can be seen from (10.11) and (10.12).

10.4 An evaluation of the license scheme[9]

From a theoretical point of view, the government can discourage improper recycling by raising the level of the fine imposed on improper recycling. However, as improper recycling in developing countries is carried out by the poor, the government cannot set the fine imposed on improper recycling at a sufficiently high level. Poor people in the informal sector know that they cannot pay the fine if improper recycling is detected, so increasing the fine fails to prevent them from improperly recycling. An enforcement policy based on a simple prohibition of improper recycling bumps up against the financial constraints of improper recyclers.

China and India, the principal E-waste recycling centers in the world, have introduced a license scheme to promote proper recycling and discourage improper recycling. One advantage of a license scheme is that it functions effectively even if the government cannot impose any fine on improper recycling, as we just saw in the preceding section. In addition, disposers of E-waste are relatively richer than the recyclers, so they can afford to pay a fine. In other words, the financial constraints of disposers are much less stringent than those of recyclers.

In view of the high profitability of authorized firms in the recycling industrial parks in China, we might conclude that the license scheme there has partly succeeded. In actual fact, though, does such a license scheme function well? The price for a PCB offered by the informal sector in China and in India is higher than that offered by licensed recyclers (see Table 9.5). According to the theoretical results looked at in Section 10.3, this price discrimination itself proves that license schemes in both countries function effectively. In China the government obliges exporting firms in developed countries to sell E-waste to licensed recyclers. Although the license scheme in India is lacking in disposer responsibility, the potential loss of reputation plays a role similar to disposer responsibility or a fine imposed on selling to the informal sector. As a result, multinational IT companies contract with licensed recyclers.

10.5 How the international trade of E-waste could be regulated

10.5.1 The disadvantages of the Basel Ban Amendment

Now we are ready to answer the final question as to how we should or could regulate the material flows of E-waste. The ban amendment of the Basel Convention was proposed as a measure to strengthen the Convention's enforcement capability. The amendment prohibits all exporting of hazardous wastes from member states of the OECD, EU, and Liechtenstein to all other countries, but it has not come into effect yet. The advantage of the Basel Ban Amendment is that the regulation applies to exporting countries. As seen in Chapter 9, regulations that apply to importing countries, such as those formulated in the Basel Convention, are not effective.

However, there are several disadvantages to the amendment. First of all, under the amendment we have to abandon low-cost recycling. Because recycling is quite labor intensive, the cost of recycling E-waste will increase dramatically when the amendment comes into effect. The US and Japan are opposed to the amendment for this reason. The US has not even ratified the Basel Convention itself.

Second, the amendment may result in the spread of improper treatment of waste not only in developing but also in developed countries, because it dramatically increases the recycling cost of end-of-life goods in developed countries. Remember that it is not only disposers but also waste management firms or recyclers who have incentives for dumping waste (see Chapter 1). Consequently, even under EPR policies, the amendment adds another incentive for recyclers to dump E-waste.

If the Basel Convention were further amended to the extent that no hazardous wastes can be traded even among non-OECD countries, the kind of improper recycling that is taking place in China, India, and Pakistan would spread to other places. If the Chinese government were to completely ban the importation of PCBs, PCBs generated in Vietnam would be recycled in Vietnam. And in fact several Chinese-funded firms have already begun to recycle PCBs in Haiphong, Mong Cai, and Ho Chi Minh City as a reaction to the strict regulation of the trade in PCBs. Thus the environmental pollution that ensues from the recycling of PCBs has already been observed in some areas of Vietnam.[10] If E-waste scrap were generated in other areas that lack recycling technologies (for example African countries), end-of-life EEE would be discarded into the environment.

Third, the most difficult problem that will confront the world after the amendment is that the amendment will lead to increasing exportation of second-hand items. The exportation of E-waste scrap will be replaced by that of second-hand items. Note that second-hand items are not considered waste, and so they are not regulated by the Convention. In that case even the amendment might not effectively regulate the transboundary movement of hazardous wastes. An increase in imported second-hand EEE makes it more difficult to collect end-of-life EEE and to transfer it to proper or licensed recyclers in developing countries.

If the Convention were also to regulate the exportation of second-hand items, it would be much easier to control the transboundary movement of hazardous wastes. But in that case the economic loss would sharply increase. The value of end-of-life durable goods mainly consists in their reuse value (see Section 7.3). In addition, the demand for second-hand items in developed countries is quite small and the market is very thin. Therefore, a complete trade ban on E-waste (including second-hand EEE) would generate a loss of the reuse values and increase the flow of resource inputs throughout the whole world.

The conclusion one comes to is that it is more desirable to establish an international recycling system that encompasses not only developing but also developed countries.

10.5.2 An alternative to the Basel Ban Amendment

We can propose an alternative international trading regulation covering E-waste. In the centers of E-waste recycling, China and India, there have been established solid nuclei of proper recycling firms. In addition, licensing systems have been established in these countries. To improve the effectiveness of a license scheme, as we saw in Section 10.3 we need to enhance the responsibility of disposers of E-waste. In this regard the hint of a new idea is contained in international E-waste trade. We can extend the licensing policy adopted in developing countries so that it is also embraced by developed countries.

In the international trade of E-waste, disposers of E-waste correspond to exporting firms in developed countries. While the imposition of an effective fine on improper recyclers in China and India is hampered by severe financial constraints, a fine on exporting firms can be set at a relatively high level. The government of an importing country, however, cannot directly impose a fine on exporting firms in other countries. Therefore, it is highly recommended that the governments of exporting countries cooperate with the governments of importing countries in imposing fines on noncompliant exporters.

That having been said, E-waste is not only imported as scrap to be recycled but it is also imported as second-hand EEE to be reused. In developing countries it is much harder to collect end-of-life EEE generated by domestic households and consign the items to authorized recyclers. Because the price of an end-of-life good is positive in developing countries, households are willing to sell them to informal recyclers who are willing to pay more than authorized recyclers. In China and India an enormous number of waste collectors visit residential areas and large firms, offering to buy end-of-life EEE, which they then resell to informal recyclers. From this point of view, the importation of second-hand EEE can be seen as causing even greater environmental pollution.

It is not efficient to impose a responsibility on each disposer. But the governments of importing countries can organize waste collectors and then impose a disposer responsibility on the members of that organization.

10.6 Discussion

We have proposed an extended license scheme as a means to regulate the international trade in E-waste. There still remain several problems in the proposed scheme. First, the proposed trading regulation assumes that developed countries export E-waste directly to the recycling centers. Even though this is the typical trade pattern, some portion of end-of-life goods in developed countries are reused in a number of developing countries and then reexported to the recycling centers, as we saw in Chapter 7. This means that we have to regulate the international trade that takes place between developing countries, which is quite difficult.

Second, it is possible that second-hand items are disposed of after the reuse period in some developing countries such as those on the western side of Africa, because they are located too far from the recycling center to export E-waste scrap there.[11] The trade of second-hand items makes it difficult to control the material flows of E-waste. If exporting countries banned the exportation of second-hand items, the material flow could possibly be regulated by the extended license system proposed in Section 10.5. A trade ban on second-hand items, however, would inevitably generate a huge economic loss both in importing countries and exporting countries.[12]

Hotta *et al.* (2008) propose another international trade policy which includes a regional coordination of EPR. In developed countries or exporting countries of second-hand items, recycling and waste disposal costs for end-of-life goods are funded by fees which either consumers or producers are required to pay under an EPR policy. A portion of the EPR fund in developed countries can be transferred to the importing countries of second-hand items. The transferred money can be used as a subsidy which ensures that end-of-life goods are recycled properly by authorized recycling firms (Hotta *et al.*, 2008).

Another policy worth considering is a selective export restriction. While some developing countries have established a recycling system, other developing countries have not. A selective export restriction would allow the exportation of E-waste (including second-hand items) only when proper recycling can be ensured in the importing country. Under such a rule some developing countries could import E-waste but others could not. This rule could raise social welfare in the world if it works effectively, because it can preserve the economic benefits derived from the E-waste trade while mitigating the environmental damages ensuing from improper recycling of E-waste. More exploration of the ramifications of such a rule is needed.

11 Conclusions

Greater economic growth always hurts the environment when it generates waste. In the “business as usual” scenario waste will not decrease, because we are not able to support the Environmental Kuznets Curve hypothesis in regard to waste generation in the waste-generating country or on the international level, using econometric analysis. This prompts requests for additional public policies to solve the hazardous waste problem. This book suggests a solution for the hazardous waste problem by presenting theoretical arguments to support the evidence gained from field studies.

First, waste is defined as a material whose price is negative. Competition in the waste market differs greatly from that in a conventional goods market. This forces us to think carefully about the market structure of waste management services. In this book we discuss problems associated with both industrial and household waste.

Industrial waste has been traded in a competitive market and treated by private waste management firms. There are three features that characterize the market for waste management. These features are: asymmetric information, the heterogeneity of the waste management firms, and noncompliance. To address these problems, governments in OECD countries have introduced various kinds of policy devices. We show that if any government chooses an appropriate level of fine and an appropriate standard of environmental care, and introduces a *manifest* requirement system, all three problematic features of the market for waste management can be addressed. As a result, the social optimum can be attained in a competitive market. An alternative optimal policy using a deposit/refund system is also discussed, and the empirical implication in Japan is provided as well.

Then we ask how household waste management services should be privatized. Traditionally, household waste has been treated by municipalities. The problem is that any signals of the cost associated with disposal and recycling cannot be transmitted to consumers or to producers. The result is that the market equilibrium results in an overproduction and overconsumption of goods that is not consistent with the social optimal level. To reduce the volume of household waste and to conserve natural resources by promoting recycling, it is important to transfer the responsibility for waste management from municipalities to consumers or producers.

But then the question arises, who should pay the disposal and recycling costs, consumers or producers? That is, which policy is better for addressing the problem, a disposal fee (DF) policy or a policy based on extended producer responsibility (EPR)? Many countries have answered the above question by introducing EPR-based policies. We discuss several policies that are consistent with EPR and look at them theoretically and through actual examples. We also look at empirical responses of households to a disposal fee and analyze the findings.

EPR is an environmental policy approach in which a producer's responsibility for a product is extended to the post-consumption stage of a product's life cycle. Two key EPR-based policies are a deposit/refund system (or a DR policy) and a product take-back requirement system (or a TB policy). These EPR-based policies are found to be equally as effective as a DF policy in the first-best economy. However, EPR-based policies are judged to be superior to a DF policy when there is a possibility of illegal disposal. The effect of a policy on Design for Environment (DfE) is also examined in the theoretical framework.

In the real economy, however, various factors prevent the first-best optimal outcome. We consider two factors, one being the transaction cost associated with a recycling subsidy (or refund) and the other being illegal disposal by the consumer. The social optimum cannot be attained when there are several sources of distortion. So we have to consider a second-best policy, and we thus move on to deal with three candidates for such a second-best policy: a DF policy, a DR policy, and a TB policy. Which of these three policies is the second-best policy will depend on the relative magnitude of the price of an end-of-life good and the marginal transaction cost associated with a recycling subsidy.

Furthermore, we consider the case where the end-of-life goods are durable goods. That is, we differentiate between *durable* and *nondurable* goods. We show that the basic result, that the social optimum can be attained by either a DF policy or EPR-based policies, cannot be applied to *durable* goods even in an economy without transaction costs. In such an economy, the first-best policy is a DF policy rather than an advance disposal fee (ADF)—a policy that is an EPR-based policy. Adoption of an ADF policy results in more consumption and disposal than is good for the social optimum. In addition, under an ADF policy the lifetime of durable goods is shorter than the social optimum. If we take transaction costs into account, however, an ADF policy could be considered the second-best in some situations, since it has some transaction cost advantages over other policies. An ADF policy can be the second-best either if net disposal cost is relatively low or if repair cost is relatively high.

Finally, we consider the case where end-of-life goods can be exported to developing countries as second-hand goods or scrap to be recycled. The problem is that domestic recycling in a developed country is confronted with the high wage rates in that country. Recycling industries are labor intensive because the recycling of materials requires complete dismantling of end-of-life goods into each different type of material. Since the 1990s end-of-life electric and electronic devices such as home appliances and personal computers, which are called

E-waste, have been exported from developed to developing countries; the problem is, improper processes used to recover metal from E-waste seriously polluted the environment in developing countries, especially China, India, and Pakistan.

The material flow of E-waste is affected by the policies adopted in both exporting countries and importing countries. First of all, the adoption of EPR-based policies in developed countries affects the material flow of E-waste. EPR-based policies are not designed to take into account the trade of end-of-life goods. An ADF policy is again not the first-best policy. It is also shown that the environment of developing countries is damaged more seriously than is consistent with the social optimum when developed countries adopt an ADF policy rather than a DF policy.

Second, the material flow of E-waste can be affected by trade regulations by importing countries. Each developing country has different types of policies to import E-waste; these policies are supported by the Basel Convention. Often, these policies are complicated by the presence of exemptions that may have been intentional, or may not have been. Using the results of our findings in detailed field surveys, we show that weak enforcement of the Basel Convention has stimulated moves by some countries to amend the Convention.

The Basel Ban Amendment prohibits the export of hazardous wastes from member states of the OECD to non-OECD countries. The enforcement of the Amendment seems stronger than that of the Basel Convention, because the Amendment is led by exporting countries.

Nevertheless, we propose an alternative international trading regulation covering E-waste. In the centers of E-waste recycling, China and India, solid nuclei of proper recycling firms have been established. In addition, licensing systems have been established in these countries. We can extend the licensing policy adopted in developing countries so that it is also embraced by developed countries.

Throughout this book waste is defined as a thing whose price is negative. One and the same thing can be waste for one person but a good for another person. A typical example of such a thing is E-waste. While end-of-life goods are waste in developed countries, they are goods or valuable materials in developing countries. Because it is easier to control the flow of goods than that of waste, end-of-life goods in developed countries may be exported to developing countries. However, both second-hand items to be reused and E-waste scrap to be recycled eventually become waste even in developing countries after they have been reused and recycled. Thus the environment in developing countries is inevitably polluted.

We have two options: the ban of E-waste trade (including second-hand items), or the control of the material flow under free trade. When the waste flow cannot be controlled in developing countries, it is rational to ban the exportation of E-waste to developing countries. We should insist on domestic recycling at a high cost in developed countries. In addition, we should abandon the reuse value of durable goods.

On the other hand, if we can manage to regulate the waste flow even in developing countries, we can obtain the benefit of free trade. As discussed in Chapter 2, the responsibility of disposers is a key driver to raise controllability of the waste flow. This is why exporting countries should not be indifferent to the E-waste problem. Future research is urgently needed to help solve this problem.

Notes

1 Introduction

1 The Basel Convention on the Control of Transboundary Movements of Hazardous Wastes and Their Disposal is usually known simply as the Basel Convention. This is an international treaty that was designed to reduce the movements of hazardous waste between nations, and specifically to prevent transfer of hazardous waste from developed to less developed countries (LDCs).
2 In 1955, Simon Kuznets postulated that economic growth and income inequalities follow an inverted U-curve. The same relation to income has been claimed for environmental quality. A long series of studies have investigated whether the EKC hypothesis is valid or not.

2 The industrial waste market

1 This section relies on the simplified version of the model presented in Shinkuma and Managi (2011).
2 Environmental care refers to how waste is treated or disposed of (e.g., whether it is preprocessed in order to mitigate harmful effects, incinerated, or recycled before final disposal, or not—that is, landfilled directly—and whether the polluted water from preprocessing and final disposal is controlled, or not).
3 The government can know the total volume of waste if an information system is constructed for collecting production and consumption data.
4 When it is assumed that $n_1 + n_2 > A > n_1$, we need two types of licenses (or standards). A more general case is analyzed in Shinkuma and Managi (2011).
5 The movement of industrial waste for legal disposal is also consistent with economic rationality. Levinson (1999a, 1999b) shows that interstate competition in hazardous waste disposal taxes has significantly affected patterns of hazardous waste disposal in the United States by decreasing shipments of waste to states enacting high taxes.
6 According to Ishiwata (2002), the most popular method is to pay as other cost rather than the price. Suppose that the real price of a waste is *minus* \$100 and the buyer of the waste asks the original owner to pay \$110 as the transport cost. Then the nominal price is \$10 and the waste becomes a good or materials to be recycled.

3 Creating the market for household waste and extended producer responsibility

1 This program is introduced in Walls (2006).
2 This program is also introduced in Walls (2006).
3 This program is also introduced in Walls (2006).

4 This program is introduced in Hosoda (2004).
5 The current fee for each home appliance is presented at www.rkc.aeha.or.jp/img/price/ryoukin2010.pdf.
6 Japanese Ministry of Environment and Japanese Ministry of Economy, Trade and Industry (2006).
7 This program is also introduced in Hosoda (2004).
8 A disposal fee program was implemented in April 2008 in Joetsu City, Niigata Prefecture. It is reported that the amount of waste thrown into rubbish bins outside convenience stores increased afterwards (http://news.livedoor.com/article/detail/3604064/). Kumamoto City in Kyushu started a similar program in October 2009—after first removing all rubbish bins from parks (*Mainichi Shimbun* (October 2, 2009)).
9 Though household waste decreased in Miyazaki City in Kyushu by 15–18 percent after introducing a disposal fee program, illegal disposals in the same city more than doubled (*Asahi Shimbun* (March 4, 2003)).

4 First-best economy: comparisons between a disposal fee policy and EPR-based policies

1 The results in this section are also implied in Dinan (1993) and Shinkuma (2000) using a partial equilibrium model.
2 DfE, or the recyclability of products, is defined in different ways by several studies. Choe and Fraser (1999) represent it as the "waste content" of the product. While Eichner and Pethig (2001) define recyclability as a portion of a product's material content that is of a particular type, Calcott and Walls (2000, 2005) define it by the cost of recycling the product.

5 The second-best policy for recycling household waste

1 This chapter is based on Shinkuma (2003).
2 Walls (2004) also stresses the importance of transaction costs in recycling markets. The second-best policy is derived in Calcott and Walls (2000, 2005).
3 In the producer take-back requirement system, the profit of firm 2 should be replaced by $\Pi_2 = (P_2 - t)q_2 - cr - g^{-1}(q_2) + \beta r$. Note that firm 2 can only choose q_2. Lagrangian Ω also should be replaced by $\Omega = U(q_1, q_2) + \lambda\{\Pi - P_1q_1 - P_2q_2 - \theta(r) - \tau(q_2 - r)\}$ instead of (5.5).

6 The durability of goods and the right recycling policy for household waste

1 This chapter is based on Shinkuma (2007).
2 The revenue from recycling used goods is implicitly included in the disposal cost (d). Namely, d represents the disposal cost net of recycling revenue. In developed countries, obsolete or defective used goods must be recycled before disposal (see the Directive on Waste Electric and Electronic Equipment, for example). They are recycled in various recycling processes and the residual is dumped. Generally, the net disposal cost is positive in developed countries. For example, the net disposal cost for a refrigerator is around \$40 in Japan (see Kawakami, 2001).
3 Suppose that $w \in [w_1, \overline{w}]$ Then "the dominated option" is dominated by Option 1. To see the reason, compare the consumer surplus (*for two periods*) associated with these two options.

$$CS^1 = w - P - s + \delta P^u + \delta(w - P - s) + \delta^2 P^u$$

$$CS^d = w - P - s + \delta(kw - (1 - q)r).$$

Note that the superscript d denotes the dominated option. It is easy to see that $CS^1 - CS^d > 0$, since any consumer with $w \in [w_1, \overline{w}]$ prefers Option 1 to Option 2 (see Figure 6.1). Similarly, supposing that, $w \in [w_2, w_1]$, we can easily see that "the dominated option" is dominated by Option 2, and so on.

4 Some of the new goods (the amount is represented by $\int_{w2}^{w1} dF$) produced at the beginning of period t are purchased as used goods by consumers taking Option 2 at the beginning of period t+1 and disposed of at the beginning of period t+2. The associated disposal cost collected at the beginning of period t is $\delta^2 d \int_{w2}^{w1} dF$. This is the first term of the numerator in (6.4).

Other new goods produced at the beginning of period t are purchased as used goods by consumers taking Option 3 at the beginning of period t+1. If those used goods work normally (the probability is q), they are disposed of at the beginning of period t+2. The associated disposal cost collected at the beginning of period t is $\delta^2 qd \int_{w3}^{w2} dF$. This is the second term of the numerator in (6.4). Conversely, if those used goods are out of order (the probability is $1-q$), they are disposed of at the beginning of period t+1. The associated disposal cost collected at the beginning of period t is $\delta(1-q)d \int_{w3}^{w2} dF$. This is the last term of the numerator in (6.4).

5 This assumption implies that the society can increase social welfare if it allocates new goods to consumers with higher WTPs. Similarly, it is desirable to preferentially repair defective used goods that are allocated to consumers with higher WTPs. Therefore, a consumer with WTP included in $[w_1, \overline{w}]$ owns a new good, and a consumer with WTP in $[w_3, w_1]$ owns a used good. Note that a used good allocated to a consumer with WTP in $[w_2, w_1]$ will be repaired if it is out of order, but a used good allocated to a consumer with WTP in $[w_3, w_2]$ will be disposed of if it is out of order.

6 If a subsidy for repair is adopted, consumers must go through various procedures in order to receive a subsidy. Each consumer must prove that the used good is defective and an adequate repair is needed.

7 A recycling subsidy is provided to consumers in the model analyzed in Chapter 5. In that model we implicitly consider *nondurable* goods and it is assumed that consumers need to prepare the goods for recycling. In that case the government needs to subsidize directly the recycling activities of consumers. As for end-of-life *durable* goods, however, no preparation of the goods is needed for recycling, and so the government can subsidize recycling and waste disposal firms rather than consumers in order to decrease transaction costs.

7 International trade of waste and environmental pollution

1 This report has been one of the most important works in this field in that it is full of important and accurate information based on original investigation and it discusses an international trade policy to solve the problem. Although this chapter and the next two chapters are based on our own field studies, we are indebted to this report for several parts of the three chapters.

2 Guiyu is located in the Chaozhou region of Guangdong Province in southern China at 23.3 N and 116.3 E. It occupies 52km^2 and has a population of about 150,000.

3 The first was carried out by Shinkuma from August 20 to 28, 2005 (in Guiyu), the second by Shinkuma with Aya Yoshida from December 15 to 25, 2005 (in Shanghai, Ningbo, Taizhou, Hangzhou, and Beijing), the third by Shinkuma from August 23 to 30, 2007 (in Delhi and Mumbai), the fourth by Shinkuma and Managi from November 17 to 23, 2007 (in Delhi, Bangalore, and Chennai), the fifth by Shinkuma and Managi from

August 24 to 30, 2008 (in Kathmandu), the sixth by Shinkuma with Keisaku Higashida from November 1 to 5, 2009 (in Sharja), and the seventh by Shinkuma with Keisaku Higashida from February 13 to 20, 2010 (in Manila).

4 Development Data Group, The World Bank, 2008. 2008 World Development Indicators Online. Washington, DC: The World Bank. Available at: http://go.worldbank.org/U0FSM7AQ40.

5 Ibid.

6 We obtained this information through interviews with several Japanese shippers of scraps. They also informed us that, even though E-waste is exported as "other ferrous scraps" from Japan to China, the same cargo may be labeled as imported "copper scrap" by the Chinese government.

7 In these tables, items whose export prices were lower than $100 were counted as second-hand ones in each Customs, because there does not exist any HS code assigned to second-hand home appliances.

8 The effects of introducing EPR policies in developed countries on international trade, the environment, and social welfare

1 This chapter is a revised version of Shinkuma (2008).

2 It is estimated that in Japan 22.87 million end-of-life home appliances were generated in 2005, and a large proportion of them (7.35 million) were exported as second-hand goods to developing countries (Japanese Ministry of Environment and Japanese Ministry of Economy, Trade and Industry, 2006).

3 Recycling is quite labor intensive, because it requires sorting all parts of an item into separate lots according to material. Therefore it is difficult to consider an alternative trade pattern whereby developed countries import scrap from developing countries.

4 Suppose that we choose quadratic functions for utility functions and recycling cost functions in both countries. Then $\partial \tilde{p} / \partial \hat{p}$ does not depend on E at all (see (8.22)).

9 The effectiveness of the Basel Convention

1 Rose-Ackerman (2006), p. 11.

2 This section is based on Section 3 of Shinkuma and Huong (2009).

3 Mong Cai (Vietnam) is located on the border between China and Vietnam. Part of the borderline is a river, and Dongxin (China) is located just across that river.

4 Manufacturers' Association for Information Technology, "First MAIT-GTZ study reveals extent of e-waste challenge," Press release, December 13, 2007. www.mait.com/admin/press_images/press13–12–2007.html

5 Ibid.

10 The policies adopted in China and India and the international trade policy

1 This section is based on Section 2 in Shinkuma and Managi (2010a).

2 Recycling industrial parks have been established in Tianjin, Taicang, Ningbo, Taizhou, and Zhangzhou.

3 This section is based on Section 3 in Shinkuma and Managi (2010a).

4 The price of E-waste is not unique under a licensing policy. The price per ton offered by a licensed recycler is represented by p_1; that offered by an unlicensed recycler is represented by p_0.

5 In China there are not a few licensed recyclers who used to be informal recyclers.

6 The detection probability of being detected operating without a license is also represented by μ, because, as stated below, unlicensed recyclers have no incentive to recycle properly and always choose improper recycling.

7 We focus on an equilibrium in which both licensed and unlicensed recyclers operate in the market, although theoretically an alternative equilibrium in which only one of them operates can exist. Suppose that equation (10.1) does not hold. For example, suppose that the right-hand side of (10.1) is strictly larger than the left-hand side. Then no disposer will sell E-waste to an unlicensed recycler, and in that case unlicensed recyclers cannot operate. This contradicts the assumption that both types of recyclers operate in the market.

8 Denote the interest rate by r. We can represent the discount factor β by $1/1 + r$.

9 This section is based on Section 4 in Shinkuma and Managi (2010a).

10 According to Yoshida and Terazono (2010), a portion of PCBs is improperly recycled at small workshops, such as those in Cavite City in the Philippines.

11 According to Schmidt (2006), while many second-hand items are imported to Nigeria, they are neither recycled nor reexported as scrap after the reuse period. They are simply disposed of at dump sites. He points out that it could be just a matter of time before the same improper metal recovery methods observed in China emerge in Nigeria.

12 In the Philippines, imported second-hand items are used for about five years (Yoshida and Terazono, 2010).

References

Basel Action Network and Silicon Valley Toxics Coalition (2002), "Exporting Harm: The High-Tech Trashing of Asia." Available via DIALOG. www.ban.org/E-waste/technotrashfinalcomp.pdf. Cited September 10, 2010.

Brock, W. and M.S. Taylor (2006), "Economic growth and the environment: A review of theory and empirics," in Durlauf S. and P. Aghion (eds.), *The Handbook of Economic Growth*, Amsterdam, Elsevier Science Publishers.

Calcott, P. and M. Walls (2000), "Can downstream waste disposal policies encourage upstream 'Design for Environment'?," *American Economic Review: Papers and Proceedings* 90: 233–237.

Calcott, P. and M. Walls (2005), "Waste, recycling, and 'Design for Environment': Roles for markets and policy instruments," *Resource and Energy Economics* 27: 287–305.

Cassing, J. and T. Kuhn (2003), "Strategic environmental policies when waste products are tradable," *Review of International Economics* 11: 495–511.

Choe, C. and I. Fraser (1998), "The economics of household waste management: A review," *The Australian Journal of Agricultural and Resource Economics* 42: 269–302.

Choe, C. and I. Fraser (1999), "An economic analysis of household waste management," *Journal of Environmental Economics and Management* 38: 234–246.

Copeland, B.R. (1991), "International trade in waste products in the presence of illegal disposal," *Journal of Environmental Economics and Management* 20: 143–162.

Copeland, B.R. (1994), "International trade and the environment: Policy reform in a polluted small open economy," *Journal of Environmental Economics and Management* 26: 44–65.

Dasgupta, S., B., Laplante, H., Wang, and D. Wheeler (2002), "Confronting the Environmental Kuznets Curve," *Journal of Economic Perspectives* 16: 147–168.

Dinan, T. (1993), "Economic efficiency effects of alternative policies for reducing waste disposal," *Journal of Environmental Economics and Management* 25: 242–256.

Eichner, T. and R. Pethig (2001), "Product design and efficient management of recycling and waste treatment," *Journal of Environmental Economics and Management* 41: 109–134.

Eichner, T. and R. Pethig (2005), "Efficient policies for green design in a vintage durable good model," *Environmental and Resource Economics* 30: 259–278.

Fullerton, D. and T.C. Kinnaman (1995), "Garbage, recycling and illicit dumping," *Journal of Environmental Economics and Management* 29: 78–91.

Fullerton, D. and T.C. Kinnaman (1996), "Household responses to pricing garbage by the bag," *American Economic Review* 86: 971–984.

Fullerton, D. and W. Wu (1998), "Policies for green design," *Journal of Environmental Economics and Management* 36: 131–148.

Grace, R., R.K., Turner, and I. Walter (1978), "Secondary materials and international trade," *Journal of Environmental Economics and Management* 5: 172–186.

Grossman, G.M. and A.B. Krueger (1995), "Economic growth and the environment," *The Quarterly Journal of Economics* 110: 353–377.

Harrington, W. (1998), "Enforcement leverage when penalties are restricted," *Journal of Public Economics* 37: 29–53.

Hastie, T.J. and R.J. Tibshirani (1990), *Generalized Additive Models*, New York, Chapman and Hall.

Hendel, I. and A. Lizzeri (1999a), "Interfering with secondary markets," *RAND Journal of Economics* 30: 1–21.

Hendel, I. and A. Lizzeri (1999b), "Adverse selection in durable goods markets," *American Economic Review* 89: 1097–1115.

Higashida, K. and N. Jinji (2006), "Strategic use of recycled content standards under international duopoly," *Journal of Environmental Economics and Management* 51: 242–257.

Hosoda, E. (1999), *The Economics of Goods and Bads* (in Japanese), Tokyo, Toyo Keizai Publishers.

Hosoda, E. (2004), "Evaluation of EPR programmes in Japan," in OECD, *Economic Aspects of Extended Producer Responsibility*, Paris, OECD.

Hotta, Y., M., Elder, and M. Tanaka (2008), "Policy considerations for establishing an environmentally sound regional material flow in East Asia," *The Journal of Environment & Development* 17: 26–50.

Huhtala, A. and E. Samakovlis (2002), "Does international harmonization of environmental policy instruments make economic sense?," *Environmental and Resource Economics* 21: 259–284.

Huo, X., L., Peng, X., Xu, L., Zheng, B.Q.Z., Qi, B., Zhang, D., Han, and P. Piao (2007), "Elevated blood lead levels of children in Guiyu, an electronic waste recycling town in China," *Environmental Health Perspectives* 115: 1113–1117.

Ishiwata, M. (2002), *The Industrial Waste Connection* (in Japanese), Tokyo, Wave Publishers.

Japan Environmental Sanitation Center, Japanese Ministry of Environment, Japanese Ministry of Economy, Trade and Industry (2007), Report on which type of second-hand EEE and E-waste scrap is regulated by Basel Convention (in Japanese).

Japanese Ministry of Environment and Japanese Ministry of Economy, Trade and Industry (2006), "The data presented in the fourth joint council on recycling of end-of-life home appliances" (in Japanese). Available via DIALOG. www.env.go.jp/council/03haiki/y0311–04/mat06.pdf. Cited September 10, 2010.

Johnstone, N. and J. Labonne (2004), "Generation of household solid waste in OECD countries: An empirical analysis using macroeconomic data," *Land Economics* 80: 529–538.

Kawakami, K. (2001), "Outline of law for recycling specified kinds of home appliances in Japan," *Environmental Economics and Policy Studies* 4: 211–218.

Kinnaman, T.C. (2009), "The economics of municipal solid waste management," *Waste Management* 29: 2615–2617.

Krutilla, K. (1991), "Environmental regulations in an open economy," *Journal of Environmental Economics and Management* 20: 127–142.

Levinson, A. (1999a), "NIMBY taxes matter: The case of state hazardous waste disposal taxes," *Journal of Public Economics* 74: 31–51.

Levinson, A. (1999b), "State taxes and interstate hazardous waste shipments," *American Economic Review* 89: 666–677.

Managi, S. (2006), "Pollution, natural resource and economic growth: An econometric analysis," *International Journal of Global Environmental* 6: 73–88.

Managi, S., A., Hibiki, and T. Tsurumi (2009), "Does trade openness improve environmental quality?," *Journal of Environmental Economics and Management* 58: 346–363.

OECD (2001), *Extended Producer Responsibility: A Guidance Manual for Governments*, Paris, OECD.

Rose-Ackerman, S. (2006), *International Handbook on the Economics of Corruption*, Cheltenham, Edward Elgar.

Schmidt, C.W. (2006), "Unfair trade: E-waste in Africa," *Environmental Health Perspectives* 114: A233–235.

Schwarzer, S., A., De Bono, G., Giuliani, S., Kluser, and P. Peduzzi (2005), "E-waste, the hidden side of IT equipment's manufacturing and use," *Environment Alert Bulletin* 5.

Shinkuma, T. (2000), "The comparisons of household waste recycling policies," *Studies in Regional Science* 30: 59–68.

Shinkuma, T. (2003), "On the second-best policy of household's waste recycling," *Environmental and Resource Economics* 24: 77–95.

Shinkuma, T. (2007), "Reconsideration of an advance disposal fee policy for end-of-life durable goods," *Journal of Environmental Economics and Management* 53: 110–121.

Shinkuma, T. (2008), "Extended Producer Responsibility in a developed country and the effect on international trade, the environment, and social welfare," Working Paper, Kansai University F-32.

Shinkuma, T. and N.T.M. Huong (2009), "The flow of E-waste material in the Asian region and a reconsideration of international trade policies on E-waste," *Environmental Impact Assessment Review* 29: 25–31.

Shinkuma, T. and S. Managi (2010a), "On the effectiveness of a license scheme for E-waste recycling: The challenge of China and India," *Environmental Impact Assessment Review* 30: 262–267.

Shinkuma, T. and S. Managi (2010b), "Enforcement leverage of a licensing system in waste management," Kansai University Working Paper, F-41.

Shinkuma, T. and S. Managi (2011), "License scheme: An optimal waste management policy under asymmetric information," *Journal of Regulatory Econoncics*, (forthcoming).

Terazono, A., A., Yoshida, and R. Murakami (Suzuki) (2007), "Summary of the regulations on trade on E-waste and waste plastics" (in Japanese), in Terazono, A., Murakami, S., Yoshida, A., Murakami (Suzuki), R., and T. Shinkuma, *Annual Report on Analysis of Material Cycle Systems for E-waste and Waste Plastics in Asia*, Tokyo, National Institute for Environmental Studies.

Terazono, A., S., Murakami, A., Yoshida, R., Murakami (Suzuki), and T. Shinkuma (2009), *The Final Report on Analysis of Material Cycle Systems for E-waste and Waste Plastics in Asia* (in Japanese), Ministry of Environment of Japan.

Tsurumi, T. and S. Managi (2010a), "Decomposition of the Environmental Kuznets Curve: Scale, technique, and composition effects," *Environmental Economics and Policy Studies* 11: 19–36.

Tsurumi, T. and S. Managi (2010b), "Does energy substitution affect carbon dioxide emissions–income relationship?," *Journal of the Japanese and International Economies* (forthcoming).

Van Beukering, P.J.H and M.N. Bouman (2001), "Empirical evidence on recycling and trade of paper and lead in developed and developing countries," *World Development* 29: 1717–1737.

Walls, M. (2004), "EPR policy goals and policy choices: What does economics tell us?", in OECD, *Economic Aspects of Extended Producer Responsibility*, Paris, OECD.

Walls, M. (2006), "Extended Producer Responsibility and product design: Economic theory and selected case studies," Resources for the Future Discussion Paper 06–08.

Widmer, R., H., Oswald-Krapf, D., Sinha-Khetriwal, M., Schnellmann, and H. Boni (2005), "Global perspectives on E-waste," *Environmental Impact Assessment Review* 25: 436–458.

Wong, M.H., S.C., Wu, M.J., Deng, X.Z., Yu, Q., Luo, A.O.W., Leung, C.S.C., Wong, W.J., Luksemburg, and A.S. Wong (2007), "Export of toxic chemicals—A review of the case of uncontrolled electronic-waste recycling," *Environmental Pollution* 149: 131–140.

Yoshida, A. (2006), "Policies on waste disposal and recycling in China" (in Japanese), in *Report on Waste Disposal and Recycling in Asian Countries*. Tokyo, Institute of Developing Economies, Japan External Trade Organization, Japanese Ministry of Economy, Trade and Industry, 19–48.

Yoshida, A. and A. Terazono (2010), "Reuse of second-hand TVs exported from Japan to the Philippines," *Waste Management* 30: 1063–1072.

Index

9780415702911